Know Your New Zealand Seashells: Hinemoana's Family

Murdoch Riley

Viking Sevenseas NZ Ltd.
P.O. Box 152, Paraparaumu,
New Zealand

Shell classifications are from "Checklist of the Recent Mollusca described from the New Zealand Exclusive Economic Zone" by H.G. Spencer, R.C. Willan, B.A. Marshall & T.J. Murray. 2002. http://toroa.ac.nz/pubs/spencer/Molluscs/index.html

Measurements of the shells listed are from "New Zealand Mollusca" by A.W.B. Powell, published by William Collins Publishers Ltd; Auckland 1979.

Portions of the introduction and other text are from "New Zealand Shells and Shellfish" by Glen Pownall, published by Sevenseas Publishing Pty. Ltd; Wellington 1971.

© 2003 Murdoch Riley
Viking Sevenseas NZ Ltd;
P.O. Box 152, Paraparaumu, New Zealand
ISBN 0854671021

Introduction

The islands of New Zealand have been isolated from any other land mass for a very long time and much of the flora and fauna of the country is native to it. This applies also to the seashells found along the long New Zealand coastline that forms a narrow band, sometimes only a few hundred metres across, where dwell the animals that produce the shells.

From the isolation of the country over eons of time has come the unique character of a large number of the endemic shells. New Zealand occupies a special and individual position in regards to the ocean currents that wash her shores and this has had a profound influence on the sea creatures that have been able to establish themselves. The illustration on the next page gives indication of the complexities of the currents on which, and in which, the ancestors of many of the sea animals came to the country.

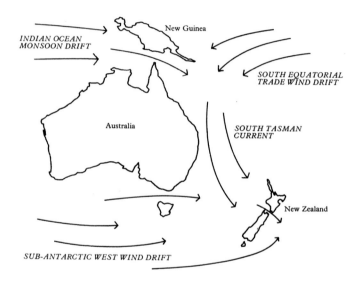

From the north, bearing slightly westwards, comes a great mass of slowly moving tropical water, driven by the ever-blowing trade winds of the Southern Hemisphere. From the south, bearing eastwards, comes the cold water of the Sub-Antarctic seas. Depending on variations of wind and tide these two great masses of slow moving water alternately lap the shores of New Zealand and offer rich food sources to the marine life of the coasts.

The westward set of the trade wind drift carries water across the South Pacific basin to flow down the east coast of Australia and continues to circulate eastward as the South Tasman current and across to New Zealand. Somewhere in its slow circulating path the waters which are destined to reach New Zealand in that very South Tasman current are joined with the outflow of the Indian Ocean which moves past Indonesia and New Guinea to reach the north-eastern area of the Australian waters.

From these varied sources have come the shells of New Zealand, but so slow is the passage of natural events that countless generations of creatures can be born and die between the arrival and establishment of one species and another. In answer to the relentless law of nature to adapt or die, each generation of newcomers tends to differ slightly from the characteristics of the first ancestor and over long periods of time species remain, which, through adaptation to local conditions, are different from species elsewhere. It is this adaptation to a new environment that has made many New Zealand shells differ from those found anywhere else in the world.

Te Marae Nui O Hinemoana
"The Vast Marae Of Hinemoana"

Hinemoana, the Ocean Maiden, is the personified form of the ocean in most accounts of Māori mythology. She is a grand-daughter of Tāne, the only god of the lower realms who was granted access to Io, the supreme god living in the uppermost heaven. Hinemoana married Kiwa, the guardian being of the ocean who produces all its waters. These are known as *Te Moana-nui-a-Kiwa*. Hinemoana's duty is to maintain the arrangement of all ocean currents and other matters. Despite being a relation of Tāne, the forest god, she assists his immortal enemy Tangaroa, fish god and tide controller, to assail the body of Papa, the Earth Mother. Indeed, all the bays and inlets of the shoreline we see today are the result of the ceaseless poundings of her legions of waves, *te ngaunga a Hinemoana*, the gnawing of Hinemoana.

Likewise Tangaroa's efforts are recognised in the proverb *he wai Tangaroa i haere ai ki uta*, "by means of water is Tangaroa enabled to go inland". Meaning that fish are able to move inland via the country's waterways, whose entrances from the sea Tangaroa has had a part in forming. Ever is Tangaroa consuming the trees and foliages of Tāne carried down to him by storm and flood. Tangaroa is sometimes given the identity of Te Parata and designation of "tide controller" because of the translation of *tanga roa* as "deep and long breather". Te Parata is known as a deep sea monster who causes low and high tide to occur by breathing vast quantities of water in and out.

The progeny of Hinemoana include species of shellfish and seaweeds. These little ones are placed under the foster care of Rakahore and Tuamatua, who are representatives of rocks and stones in Māori belief. They provide shelter for Hinemoana's offspring. Some species of shellfish are given to be the progeny of Te Arawaru and Kaumaihi. These include Pipi and Kuku, personifications of cockles and mussels respectively. They and other sea creatures are known as *te whānau a Te Arawaru*.

Long, long, ago there was war between the Pipi and Kuku tribes. In 1929 Elsdon Best described this early episode in his book "Fishing Methods And Devices Of The Māori":

"An old folk-tale describes a quarrel that arose in remote times between the Kuku and Pipi families at Waikaru – that is, between mussels and cockles. This strife raged at Onetahua, where the cockles dug themselves in; the sand-beaches were their defensive works. The mussel folk attacked them. When they thrust out their "tongues" they became clogged with sand, and so the mussel folk were defeated by the cockle tribe, hence the latter are still seen holding the sandy beaches, while the mussels have to cling to Rakahore – that is, rocks off shore. Ever does Hineone, the Sand Maid, shelter the offspring of Te Arawaru.

Now, when Te Pūwhakahara and Takaaho heard of this quarrel they said, "What are these small folk quarrelling about?" Said the former to Takaaho, "Our offspring are in want of food: procure the progeny of Te Arawaru as food for them". (These offspring of Takaaho and Te Pūwhakahara are the various species of sharks and whales). Said Takaaho, "But they will retire behind their sand breastworks and so defy us". Te Pū replied, "Scoop them up as they appear from their breastworks". So Takaaho went forth with a band to Onetahua, whereupon the offspring of Te Arawaru, the cockles, fled to their fortified village – that is, they burrowed into the sand. The party of Takaaho then delivered an attack; but these folk were defeated – their gills filled with sand; and this is why we see whales perish on sandy beaches".

Kai Mātaitai:
Food Of The Salt Sea

Certain rituals applied to fishing, rather fewer to shellfish collection. There were rules as to how and when shellfish could be taken. There were temporary restrictions known as *rāhui* placed on shellfish beds during the breeding seasons, or if a chief had died, or a person had been drowned close by. Women were not allowed to join in any food gathering during their menstrual periods for fear the shellfish would migrate further along the beach. It was also a gross offence to fish or collect shellfish on grounds that belonged to another people.

Many different charms were recited to make a fishing expedition or shellfish collection successful. Some were of extraordinary complexity, and some are still in use. One of most direct is this one from the papers of John White (MS Papers 75, B35/2, Alexander Turnbull Library), with modern translation: -

Te patua koe i uta	You can be killed on land
Te patua koe i tai	You can be killed at sea
Te patua Tāne	Tāne can be killed
Te patua Tangaroa	Tangaroa can be killed

He kaha koe i uta	Have strength on land
He kaha koe i tai	Have strength at sea
He kaha Tāne	Be strong Tāne
He kaha Tangaroa	Be strong Tangaroa

Elsdon Best in "Fishing Methods And Devices Of The Māori" described the division of labour between men and women: "Fishing was viewed by the Māori as being essentially a task for men, and so to them most of the work was left. Probably the *tapu* pertaining to the art, especially sea-fishing from a canoe, had some effect in preventing women taking part in this highly useful employment. Women, however, were the collectors of shellfish, and also took part in taking small fresh-water fish".

Men assisted with the more difficult work, such as the job of prising pāua off rocks at depth. An implement called a *ripi* was used. A famous episode from Ngāti Kahungunu history relates how their ancestral father, Kahungunu, showed his prowess. In diving for pāua he held his breath longer than others of the village he was visiting could do, and he dived deeper than others had previously been able to do. Thus he gave proof that he would be an able provider for the tribe, and incidentally this helped him win the love of his desired one, Rongomaiwahine.

It is a common saying that there is nothing new under the sun, and indeed this is true. Today there are many marine farms that cultivate both oysters and mussels for the market, and they range from Northland to Stewart Island. Yet long ago the Māori had developed a method of farming these self-same seafood delicacies, oysters (tio) and mussels (kuku). John White recorded this information in Northland in 1848:

"The ancient Māori when they intended to make a feast at which the kuku mussel shellfish was to be one of the delicacies (grew their own supplies). Some months before the feast was to take place they cut down a number of trees called Tīpau (red matipo) and put them in the river near the sea coast. These they stuck up at low water tide and left them. On these in about four or six months there would be mussels the size of a man's fist".

"(Also) the twigs and large branches of the tree called Poporo-kai-whiria (pigeon wood) are taken and stuck up on a sand bank where the water is about the depth of eighteen inches below low water mark. This is done in a long line all along the bank. To these branches the mussels cling and grow large in about two months. Such are very fast and attain the size of double those that adhere to rocks. Oysters are also propagated and feed in the same manner. To have a continuous supply, new trees of the same kind are put in as the old ones decay. After a year thus growing, the mussels in a space of half an acre will keep a tribe of say one hundred persons with mussels all the year". (John White, MS Papers 75/B36/56 and 75/143, Alexander Turnbull Library, Wellington).

Rou kākahi:
Fresh-water mussel rake
(Frame without basket)

Rongoa No Te Tai:
Medicines From The Sea

Shellfish remedies were used by the Māori to treat and cure, much as occurred in plant medicine. In both cases the belief was that medicine would only work when it went hand in hand with spirituality. At Stewart Island the soft roe of the limpet called kākahi ngākihi was given as an opening medicine to newly born infants. In Canterbury seal fat served the same purpose. In Northland the juice of the pipi was given to invalids and to those who had just given birth to a child. This juice when cooked with the leaves of wild turnip (nanī) was said to cure colds, coughs or sore throat. See "Māori Healing and Herbal" by Murdoch Riley for more details on medical practices.

Seaweeds too were much used. Soft globular varieties like "Neptune's necklace" were chewed for sore throats and for chest infections, the juice also used on sore ears. The whiplike ends of bull kelp were an aid to expelling worms, it is said, and "to ward off" goitre and rheumatics. The berry juices of the tutu plant (less the seeds) were fermented with the dried seaweed called karengo and used to counteract illness caused through eating tōtara, rimu or karaka berries. Poultices of seaweed were used on swollen joints. In one case a hot pack of seaweed was used on a patient to alleviate the effects of concussion. Seaweed that had been gathered at high tide, and while still impregnated with sea water, was used effectively as a pack on katipō spider bites. Seal oil, whale oil, shark oil and meat, stingray oil, and the roe of mullet fish have all been used for medicinal purposes.

A Māori Love Story:
"The Murmur Of The Shell"

This romantic tale was recorded by Captain Gilbert Mair and first published in an Adelaide newspaper in 1867. It is written in the flowery style of the time and is reproduced, in slightly condensed form, from the Auckland Star daily of August 12th 1922.

In the old days the Māori were wont to hold functions called Whakahorohoro, a kind of military tournament where warriors from far and near would foregather to compete in warlike exercises, games of skill and strength such as the ancient Greeks indulged in on "High Olympics". During the progress of these tournaments a truce existed, and it was considered the highest etiquette to bestow marked hospitality toward unfriendly tribes, for the old time Māori was not wanting in chivalry.

It was at such a function, given by the powerful Whakatohea tribe of Ōpōtiki, that a young chief from Turanganui named Tawhito, excelled all others in comeliness and trials of strength and he met Tauputaputa (translates as "the changing year") who was the pride and beauty of her tribe. But he was a modest youth and concealed his feelings till the days of grace had expired, when he besought her parents to let the fair Tauputaputa become his wife, but they strongly refused an alliance with the Gisborne people on account of long-standing animosities.

Tauputaputa, on her part, felt strangely drawn towards the young chief, but she was shy and diffident. Tawhito sadly returned to his home, and a year passed filled with hopeless longings, till he thought of how he should make the loved one return his passion. He consulted the family priest who instructed him what to do. He went to the seashore, and taking a toitoi shell, he breathed into it a love incantation, or spell (a *whakaahu wahine*), then cast it from him. This charm he repeated four times, throwing the shell to the four winds.

On the morrow his yearning was so great that he determined to seek her who possessed his soul, and for her sake to brave the hostility of her tribe and the danger of passing through enemy country. He informed no one of his intentions, and one night when his people slept, he started forth up the Waipaoa River, and by the next night he had entered the great forest which extends toward Ōpōtiki, where, for the present, we shall leave him.

Tauputaputa sighed in hopeless despair for her noble young warrior and her parents became solicitous for her health, and encouraged the village maidens to come and join in the amusements. One day her mother persuaded them all to sally forth on Awakino Beach, promising that the girl who found the largest toitoi would win a handsome husband. Soon they espied such a shell washed up by the waves, and raced for it, but Tauputaputa, being swiftest of foot, outstripped the others, and on finding the shell dead with a hole in it, she cast it away in disdain. But that dead toitoi shell would not be denied and thrice it was cast up at her feet, only to be thrown back in anger far into the sea.

At last, when the shell appeared for the fourth time, she thought: "Surely this is more than a coincidence". So she went apart from her companions, and taking the shell with her, sat on a rock, when the memory of Tawhito filled her mind with pain and she wept over the hopelessness of her love. Bowing her head on her lap her ear touched the shell which seemed to say: "Tawhito sent me to tell of his love". Hope was renewed, and she gathered leaves of the sweet-scented kāretu grass, weaving a string, and hung the messenger shell around her neck.

For many nights she dreamed of Tawhito and she determined to seek him, first of all making a confidant of her father's tohunga, to whom she told her secret about the shell. He said it was a favourable omen; that it meant that Tawhito was on his way to meet her, and she must wait. But she became terrified lest he fall into the hands of his enemies, which would mean certain death, therefore she must go at once.

The tohunga endeavoured in vain to dissuade Tauputaputa, but finding that she was determined, quoted a proverb to the effect that she should not let any small obstacles turn her from her bold project. He also initiated her into a spell or invocation for success, commencing: "*I hira mata tao taonga ki te whai ao*". "Open my eyes that they may behold the precious treasure of life".

That night she heard the voice of the shell telling her, more consistently than ever, that Tawhito loved her. Though a great storm was raging, the girl sought the war trail leading to Turanganui, taking a small basket of kao, dried kūmara, tied to her waist. At nightfall on the second day she had reached Motu, Te Waonui-a-Tāne (the great shade of Tāne Mahuta's forest).

She had become hopelessly lost, when suddenly she remembered the counsel of her father's tohunga and repeated the spell he had taught her. Instantly hope returned to her heart and the blood coursed rapidly through her veins. Tauputaputa sprang to her feet groping her way carefully through the dense undergrowth, when, feeling a smooth surface under her, she discovered she had regained the lost trail, which she followed for some distance till she smelled the smoke of a tawa fire and soon discerned a faint glimmering light.

Then a great fear came upon her lest she fall into an enemy's hands and a sad fate await her. Cautiously crawling on her hands and knees she drew nearer and espied a solitary figure bowed over the fire. Emboldened, she approached nearer, and who can tell her joy at meeting Tawhito!

Here the curtain falls and when it is raised again Tawhito and his wife Tauputaputa are living happily in his pā at Titirangi on the hill above the entrance to Turanganui River. And she bore him many children whose offspring live on the coast at various places between Māhia and Uawa; and their maidens to this day, when playing on the seashore, still search for toitoi shells which they press to their ears hoping to hear the message of love which came bringing joy to Tauputaputa of the changing year.

Seashell Groups

All true seashells are the exoskeletons of a group of animals known as the phylum Mollusca. The number of molluscs living today in the world is given as somewhere around 100,000 species, with around 2500 known to live in New Zealand. The mollusca derive their support and refuge within an external skeleton in the form of a shell that is deposited by the animal from the living cells of its mantle. This shell consists mainly of calcium carbonate or limestone, usually in three distinct layers. In some types of shells the age of the animal at the time it lost its shell can be determined by the annual growth rings displayed on the shell. The inner layer is made of nacre, mother-of-pearl.

Mollusca are divided into seven groups, five of which, the principal ones, are illustrated in this book in taxonomic order. These are: -

1. **Chitons** (Class Polyplacophora) (Page 28)
    Also called coat-of-mail shells for their eight

plates, usually overlapping, firmly held in place by a surrounding fleshy girdle. A primitive class dating back to pre-historic times.

2. **Univalves** (Class Gastropoda) (Pages 28 -50)

By far, the largest class with about 90,000 living species worldwide. The shell is in one piece, the opening to which can in some species be closed from the inside by a shelly cap called the operculum. They come in many forms but spiralled type is the most common.

3. **Tusk Shells** (Class Scaphopoda) (Page 51)

Rather rare deep sea animals, not usually seen alive. Their shells are slightly curved with a gradually tapering form open at both ends. They are marine burrowing molluscs and as a rule allow only their narrow posterior extremity to project from the sand in which they hide.

4. **Bivalves** (Class Bivalvia) (Pages 51-68)

Another large class. The shell is in two pieces connected by a hinge (or valve), the animal using its strong muscles to open and close them. For example pipi, mussels and oysters are bivalves. After death of the animal the two valves often become separated or spread out.

5. **Cephalopoda** (Class Cephalopoda) (Page 68)

To this class belong all those sea creatures with long sucker-bearing arms including the octopus, the cuttlefish and the squid. Cephalopods usually have a very much-modified shell, or none at all.

Dwellers Of Mud Flats

- 14 Tiara Top Shell (*Trochus tiaratus*)
- 19 Mudflat Top Shell (*Diloma subrostrata*)
- 32 Horn Shell (*Zeacumantus lutulentus*)
- 33 Turret Shell (*Maoricolpus roseus*)
- 43 Ostrich Foot (*Struthiolaria papulosa*)
- 56 Common Trophon (*Xymene plebeius*)
- 67 Speckled Whelk (*Cominella adspersa*)
- 68 Spotted Whelk (*Cominella maculosa*)
- 70 Mud Whelk (*Cominella glandiformis*)
- 76 Arabic Volute (*Alcithoe arabica*)
- 84 Brown Bubble Shell (*Bulla quoyi*)
- 85 White Bubble Shell (*Haminoea zelandiae*)
- 86 Mud Snail (*Amphibola crenata*)
- 89 Tusk Shell (*Antalis nana*)
- 90 Razor Mussel (*Solemya parkinsonii*)
- 91 Mallet Shell (*Neilo australis*)
- 102 Horse Mussel (*Atrina zelandica*)
- 119 Silky Dosinia (*Dosinia lambata*)
- 123 Coarse Dosina (*Dosina zelandica*)
- 126 Cockle (*Austrovenus stutchburyi*)
- 128 Oblong Venus Shell (*Ruditapes largillierti*)
- 131 Large Wedge Shell (*Macomona liliana*)
- 132 Round Wedge Shell (*Pseudocopagia disculus*)
- 137 Oval Trough Shell (*Cyclomactra ovata*)
- 142 Pipi (*Paphies australis*)
- 148 Lantern Shell (*Offadesma angasi*)

Dwellers Of Rocky Shores

1. Butterfly Chiton (*Cryptoconchus porosus*)
2. Green Chiton (*Chiton glaucus*)
3. Pāua (*Haliotis iris*)
4. Silver Pāua (*Haliotis australis*)
5. Virgin Pāua (*Haliotis virginea*)
6. Shield Shell (*Scutus breviculus*)
7. Black Edged Limpet (*Notoacmea pileopsis*)
8. Denticulate Limpet (*Cellana denticulata*)
9. Radiata Limpet (*Cellana radians*)
10. Golden Limpet (*Cellana flava*)
11. Star Limpet (*Cellana stellifera*)
12. Ornate Limpet (*Cellana ornata*)
13. Southern Limpet (*Cellana strigilis redimiculum*)
15. Green Top Shell (*Trochus viridis*)
16. Dark Top Shell (*Melagraphia aethiops*)
17. Knobbed Top Shell (*Diloma bicanaliculata*)
22. Tiger Shell (*Calliostoma tigris*)
24. Spotted Tiger Shell (*Calliostoma punctulatum*)
26. Cat's Eye (*Turbo smaragdus*)
27. Granose Turban (*Modelia granosa*)
29. Cook's Turban (*Cookia sulcata*)
30. Black Nerita (*Nerita atramentosa melanotragus*)
31. Periwinkle (*Nodilittorina cincta*)
33. Turret Shell (*Maoricolpus roseus*)
38. Circular Slipper Shell (*Sigapatella novaezelandiae*)
39. White Slipper Shell (*Crepidula monoxyla*)
40. Ribbed Slipper Shell (*Crepidula costata*)
48. Spengler's Trumpet (*Cabestana spengleri*)
49. Hairy Triton (*Cymatium parthenopeum*)
50. Trumpet Shell (*Charonia lampas rubicunda*)
51. Swollen Trumpet (*Argobuccinum pustulosum tumidum*)
52. Australian Triton (*Ranella australasia*)
54. Octagonal Murex (*Murexsul octagonus*)

56 Rock Trophon (*Paratrophon patens*)
57 Stanger's Trophon (*Paratrophon quoyi*)
58 White Rock Shell (*Dicathais orbita*)
59 Dark Rock Shell (*Haustrum haustorium*)
60 Oyster Borer (*Lepsiella scobina*)
61 Many-Lined Whelk (*Buccinulum linea*)
62 Siphon Whelk (*Penion sulcatus*)
65 Southern Siphon Whelk (*Penion mandarina*)
70 Red-Mouthed Whelk (*Cominella virgata*)
84 Small Siphon Limpet (*Siphonaria australis*)
85 Large Siphon Limpet (*Benhamina obliquata*)
89 Ark Shell (*Barbatia novaezealandiae*)
95 Blue Mussel (*Mytilus galloprovincialis*)
96 Green Mussel (*Perna canaliculus*)
97 Ribbed Mussel (*Aulacomya atra maoriana*)
98 Nesting Mussel (*Modiolarca impacta*)
99 Date Mussel (*Zelithophaga truncata*)
100 Hairy Mussel (*Modiolus areolatus*)
101 Small Black Mussel (*Xenostrobus pulex*)
104 Fan Scallop (*Talochlamys zelandiae*)
108 Striped Fan Scallop (*Talochlamys zeelandona*)
112 Golden Oyster (*Anomia trigonopsis*)
113 Stewart Island Oyster (*Ostrea chilensis*)
114 Auckland Rock Oyster (*Saccostrea glomerata*)
115 Nestling Cockle (*Cardita aoteana*)
127 Ribbed Venus Shell (*Protothaca crassicosta*)
128 Oblong Venus Shell (*Ruditapes largillierti*)

Dwellers Of Sandy Shores

23 Pale Tiger Shell (*Calliostoma selectum*)
24 Wheel Shell (*Zethalia zelandica*)
33 Turret Shell (*Maoricolpus roseus*)
34 Pagoda Turret Shell (*Zeacolpus pagoda*)

35 Stewart Island Turret Shell (*Zeacolpus symmetricus*)
36 Wentletrap (*Cirsotrema zelebori*)
42 Small Ostrich Foot (*Pelicaria vermis*)
43 Ostrich Foot (*Struthiolaria papulosa*)
45 Necklace Shell (*Tanea zelandica*)
46 Helmet Shell (*Semicassis pyrum*)
55 Large Trophon (*Xymene ambiguus*)
66 Knobbed Whelk (*Austrofusus glans*)
67 Quoy's Whelk (*Cominella quoyana*)
71 Stewart Island Whelk (*Cominella nassoides*)
74 Brown Olive (*Amalda mucronata*)
75 Southern Olive (*Amalda australis*)
76 Arabic Volute (*Alcithoe arabica*)
77 Southern Volute (*Alcithoe swainsoni*)
78 Depressed Volute (*Alcithoe arabica depressa*)
80 Little Volute (*Alcithoe fusus*)
82 Pink Tower Shell (*Phenatoma rosea*)
83 New Zealand Auger (*Pervicacia tristis*)
89 Tusk Shell (*Antalis nana*)
94 Large Dog Cockle (*Tucetona laticostata*)
103 Queen Scallop (*Pecten novaezelandiae*)
111 Little File Shell (*Limatula maoria*)
116 Purple Cockle (*Purpurocardia purpurata*)
117 Lace Cockle (*Divaricella huttoniana*)
118 Strawberry Cockle (*Pratulum pulchellum*)
120 Ringed Dosinia (*Dosinia anus*)
121 Fine Dosinia (*Dosinia subrosea*)
124 Morning Star (*Tawera spissa*)
125 Frilled Venus Shell (*Bassina yatei*)
128 Oblong Venus Shell (*Notocallista multistriata*)
129 Angled Wedge Shell (*Peronaea gaimardi*)
130 Spencer's Wedge Shell (*Rexithaerus spenceri*)
133 Purple Sunset Shell (*Gari stangeri*)
134 Pink Sunset Shell (*Gari lineolata*)
135 Shining Sunset Shell (*Soletellina nitida*)

136 Large Trough Shell (*Mactra discors*)
138 Elongated Mactra (*Oxyperus elongata*)
139 Triangle Shell (*Spisula aequilatera*)
140 Scimitar Mactra (*Zenatia acinaces*)
141 Lance Mactra (*Resania lanceolata*)
142 Pipi (*Paphies australis*)
143 Toheroa (*Paphies ventricosa*)
144 Tuatua (*Paphies subtriangulata*)
145 Deep Burrower (*Panopea zelandica*)
147 Battleaxe (*Myadora striata*)

Dwellers In Deep Water

1   Butterfly Chiton (*Cryptoconchus porosus*)
25 Spotted Tiger Shell (*Calliostoma punctulatum*)
33 Turret Shell (*Maoricolpus roseus*)
37 Violet Snail (*Janthina janthina*)
39 White Slipper Shell (*Crepidula monoxyla*)
41 Carrier Shell (*Xenophora neozelanica*)
44 New Zealand Cowry (*Trivia merces*)
47 Cask Shell (*Tonna cerevisina*)
48 Spengler's Trumpet (*Cabestana spengleri*)
49 Hairy Triton (*Cymatium parthenopeum*)
50 Trumpet Shell (*Charonia lampas rubicunda*)
51 Swollen Triton (*Argobuccinum pustulosum tumidum*)
53 Spiny Murex (*Poirieria zelandica*)
54 Octagonal Murex (*Murexsul octagonus*)
64 Deep Water Siphon Whelk (*Penion cuvierianus*)
66 Knobbed Whelk (*Austrofusus glans*)
67 Speckled Whelk (*Cominella adspersa*)
69 Quoy's Whelk (*Cominella quoyana*)
72 Stewart Island Whelk (*Cominella nassoides*)
73 Pagoda Shell (*Coluzea spiralis*)
77 Southern Volute (*Alcithoe swainsoni*)

79 Calva Volute (*Alcithoe calva*)
81 Golden Volute (*Provocator mirabilis*)
89 Tusk Shell (*Antalis nana*)
91 Mallet Shell (*Neilo australis*)
92 Ark Shell (*Barbatia novaezelandiae*)
93 Small Dog Cockle (*Glycymeris modesta*)
96 Green Mussel (*Perna canaliculus*)
98 Nesting Mussel (*Modiolarca impacta*)
99 Date Mussel (*Zelithophaga truncata*)
100 Hairy Mussel (*Modiolus areolatus*)
104 Fan Scallop (*Talochlamys zelandiae*)
105 Southern Fan Scallop (*Zygochlamys delicatula*)
106 Deep Water Fan Scallop (*Talochlamys gemmulata*)
107 Purple Fan Scallop (*Talochlamys gemmulata radiata*)
108 Striped Fan Scallop (*Talochlamys zeelandona*)
109 Lion's Paw (*Mesopeplum convexum*)
110 File Shell (*Lima zelandica*)
111 Little File Shell (*Limatula maoria*)
113 Stewart Island Oyster (*Ostrea chilensis*)
115 Nestling Cockle (*Cardita aoteana*)
122 Deep Water Venus Shell (*Notocallista multistriata*)
123 Coarse Dosina (*Dosina zelandica*)
124 Morning Star (*Tawera spissa*)
149 Paper Nautilus (*Argonauta nodoso*)
150 Ram's Horn (*Spirula spirula*)

In Sponges

105 Southern Fan Scallop (*Zygochlamys delicatula*)

On Seaweed

20 Opal Top Shell (*Cantharidus opalus*)
21 Pink Top Shell (*Cantharidus purpureus*)

On Decaying Kelp

18 Blue Top Shell (*Diloma nigerrima*)

Pelagic (Inhabiting Open Oceans)

37 Violet Snail (*Janthina janthina*)

On Stones Or Shells

38 Circular Slipper Shell (*Sigapatella novaezelandiae*)

On Apertures Of Dead Univalves

39 White Slipper Shell (*Crepidula monoxyla*)

CHITONS

1. Butterfly Chiton/ Papapiko
Cryptoconchus porosus
(Family Acanthochitonaidae)

Environment: Low tidal rocks and deeper water. *Range*: New Zealand. *Colour*: Light blue-green, rarely pink. *Occurrence*: Fairly common. *Length*: 38-73mm. *Width*: 20-33mm (dried state). ***REMARKS***: Has eight internal shell valves embedded in the body of the brown-green-orange-coloured animal. A flexible girdle surrounds the valves and holds them together. The nineteenth century naturalist Quoy was given "Karimoan" as the Māori name of this chiton. "Moana" is the word for "ocean".

2. Green Chiton/ Papatua *Chiton glaucus*
(Family Chitonidae)

Environment: Under stones. *Range*: New Zealand. *Colour*: Dark green to brown, sometimes with light yellow or red-brown markings. More colourful where they have been protected from strong sunlight. *Occurrence*: Very common. *Length*: 32-55mm. *Width*: 21-35mm. ***REMARKS***: The chiton spends most of its time clasping firmly to rocks, but is capable of moving short distances with its muscular foot in search of algae and seaweed fragments.

UNIVALVES

3. Pāua *Haliotis iris*
(Family Haliotidae)
Also known as Black Footed Pāua, Mutton-fish, Ormer, Sea-Ear

Environment: In rock crevices at extreme low tide. *Range*: New Zealand. *Colour*: Exterior: Pale brown or olive-green, usually heavily encrusted. Interior: Iridescent blue-green with flashes of blue, red and green. *Occurrence*: Common. *Length*: 117mm (average). *Width*: 86mm (average). ***REMARKS***: Widely used today for jewellery and ornaments. Used by the Māori as "eyes" to inlay carvings, the eyes being described as *muri ahi*, "a blaze of fire". Pāua shell decorated the taiaha, a weapon of some one and a half metres long. Thus the saying *he kanohi taiaha*, "a taiaha with many eyes" to describe a two-faced, deceitful person. Another attribute of the pāua is its ability to adhere to rocks, even in the fiercest of storms. Hence this saying *he pāua piri toka*, "a rock-clinging pāua", used to describe someone who is absolutely dependable. Pāua is a popular food, though the taking of the shellfish is restricted. Commercial cultivation is also now carried out. In former times, when not required at once by the Māori, pāua were preserved for winter consumption. They were dried over the smoke of a fire and threaded onto flax strings. The shells were used as containers for liquids. The breathing holes were first stopped up, then used to contain water, plant oils and gums, dyes etc. Fish hooks and spinners were sometimes made from this shell.

4. Silver Pāua *Haliotis australis* (Family Haliotidae)
Other Māori names: Hihiwa, Karahiwa, Karariwha, Korohiwa, Kororiwha, Koeo and Marariwha.

Environment: Under rocks at extreme low tide. ***Range***: New Zealand. ***Colour***: Exterior: Khaki. Interior: Iridescent silver, usually tinged with pink. ***Occurrence***: Fairly common. ***Length***: 82mm (average). ***Width***: 56mm (average). ***REMARKS***: Animal is black with a foot varying in colour from white through to orange. This is distinct from the black animal pāua that has a black foot. The foot helps the animal attach itself to rocks, and also to move its home.

5. Virgin Pāua/Marapeka *Haliotis virginea* (Family Haliotidae)

Environment: Under stones at extreme low tide. ***Range***: South Island and southern part of North Island. ***Colour***: Exterior: Light brown, dotted and splashed with every conceivable colour. Interior: Iridescent green and pink. ***Occurrence***: Uncommon. ***Length***: 50mm (average). ***Width***: 32.5mm (average). ***REMARKS***: Replaced by three sub-species: *H.v. crispata* in the northern part of the North Island. *H.v. morioria* at the Chatham Islands. *H.v. huttoni* at the Sub-antarctic islands.

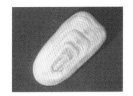

6. Shield Shell/Ngākahi *Scutus breviculus* (Family Fissurellidae)
Other Māori names: Ngākihi, Pūpū

Environment: In rock crevices at extreme low tide. ***Range***: New Zealand. ***Colour***: White. ***Occurrence***: Fairly common. ***Length***: 50-70mm. ***Width***: 28-39mm. ***REMARKS***: Shell is internal in the back of a large black slug. The shell is small compared to the animal's size. Because of its shape, it is sometimes called "duck-billed". A compliment of the 19th century was to compare the well-tattooed lips of a Māori woman with the plump black and glossy mantle of this shellfish.

7. Black Edged Limpet/Rori *Notoacmea pileopsis* (Family Lottidae)
Other Māori names: Ngākihi, Pūpū

Environment: High tidal rocks on exposed coasts. ***Range***: North Island and northern part of South Island. ***Colour***: Exterior: Dark grey with lighter mottlings. Interior: Brown in the centre, ringed by white, and edged with black. ***Occurrence***: Common. ***Length***: 19.5-33mm. ***Width***: 16-28mm. ***REMARKS***: The shell's edge is thin and sharp. Replaced in the southern part of the South Island by *N. pileopsis sturnus*.

8. Denticulate Limpet/Ngākihi *Cellana denticulata*
(Family Nacellidae)
Other Māori names: Pūpū, Rori

Environment: On intertidal rocks. *Range*: North Island and northern part of South Island. *Colour*: Exterior: Dull dark grey. Interior: Mottled black and grey, centre black. *Occurrence*: Common, but local. *Length*: 37-70mm. *Width*: 29-60mm. *REMARKS*: Strongly ribbed with many concentric ridges that thicken into granular scales.

9. Radiata Limpet/Ngākihi *Cellana radians*
(Family Nacellidae)
Other Māori names: Pūpū, Rori

Environment: On intertidal rocks. *Range*: New Zealand. *Colour*: Exterior: Variable greenish-grey with brown rays. Interior: Silvery or yellowish, rayed with black and splashed with brown. Patterns lost with age to become silvery-white. *Occurrence*: Very common. *Length*: 34-54mm. *Width*: 28.5-46mm. *REMARKS*: The form *perana* has shining yellow-green interior, with a central patch of pale grey.

10. Golden Limpet/Ngākihi *Cellana flava*
(Family Nacellidae)
Other Māori names: Pūpū, Rori

Environment: On intertidal rocks. *Range*: East Cape to North Canterbury. *Colour*: Exterior: Dull grey to dull orange. Interior: Light yellow to bright orange. *Occurrence*: Common, but local. *Length*: 36.5-66mm. *Width*: 28-54mm. *REMARKS*: A very attractive shell. Small specimens are usually more highly coloured than larger ones. Very large examples are found at Cape Campbell.

11. Star Limpet/Ngākihi *Cellana stellifera*
(Family Nacellidae)
Other Māori names: Pūpū, Rori

Environment: On rocks at extreme low tide. *Range*: North Island and northern part of South Island. *Colour*: Exterior: Dark brown. Interior: Bluish-purple/grey, with a white star in the centre. *Occurrence*: Common, but local. *Length*: 35-70mm. *Width*: 29-58mm. *REMARKS*: Live specimens are usually very encrusted on the outer surface. Beach specimens are red-brown, with a white central star. The word "*stellifera*" because the periostracum is in the form of a star. In the sub-species *phymatius* the rays of the star come right to the edge of the shell.

12. Ornate Limpet/ Ngākihi *Cellana ornata*
(Family Nacellidae)
Other Māori names: Pūpū, Rori

Environment: On intertidal rocks. *Range*: New Zealand. *Colour*: Exterior: Light brown. Interior: Black and white rays on a light brown background with a pale patch in the centre. *Occurrence*: Very common. *Length*: 23-50mm. *Width*: 18-42mm. *REMARKS*: Has eleven sharp radial black ribs, each marked with a series of white-toothed bumps. Animals living lower down on the shore are larger than those at higher levels.

13. Southern Limpet/Ngākihi *Cellana strigilis redimiculum*
(Family Nacellidae)
Other Māori names: Pūpū, Rori

Environment: On intertidal rocks. *Range*: Southern part of South Island and Stewart Island. *Colour*: Exterior: Orange-brown. Interior: Rayed white with a light brown central patch. *Occurrence*: Common. **Length:** 42-78mm. **Width:** 34-65mm. *REMARKS*: The sub-species *C. strigilis chathamensis* is confined to the Chatham Islands, and *C. strigilis strigilis* to the sub-antarctic islands.

14. Tiara Top Shell/ Mitimiti
Trochus tiaratus
(Family Trochidae)
Other Māori name: Mimiti

Environment: In sandy mud below low tide. *Range*: New Zealand. *Colour*: Whitish with tessellations of reddish-brown. *Occurrence*: Common. *Height*: 12-13.7mm. *Width*: 14-16mm. *REMARKS*: The operculum of top shells vary in colour from yellow to dark brown. The animals feed on seaweeds. The word "*tiaratus*" means "like a tiara".

15. Green Top Shell/ Tihipu *Trochus viridus*
(Family Trochidae)
Other Māori name: Tihi

Environment: On rocks at extreme low tide as well as in surf. *Range*: New Zealand. *Colour*: Exterior: Brownish to greenish, with a flat, pale yellow base. Interior: Iridescent. *Occurrence*: Fairly common. *Height*: 18-25mm. *Width*: 21-30mm. *REMARKS*: The sculpture of the shell surface consists of whorls of rounded bead-like granules. The word "*viridis*" has the meaning here of "green".

16. Dark Top Shell/Pūpūmai
Melagraphia aethiops
(Family Trochidae)
Other Māori names: Māihi, Pūpū

Environment: On intertidal rocks. *Range*: New Zealand. *Colour*: Dull olive to dark bluish-brown. Speckled with white on the spire. The white aperture is heavily calloused. *Occurrence*: Very common. *Height*: 21-28mm. *Width*: 21-31mm. *REMARKS*: A specimen of this shell was taken to England by the explorer Captain Cook in the late 18th century. The word "*aethiops*" originally referred to Ethiopians, or dark-skinned people living south of Egypt. Here simply dark-shaded. Also sometimes given as "*aethiopissa*", "the negress with the white teeth".

17. Knobbed Top Shell
Diloma bicanaliculata
(Family Trochidae)

Environment: Under intertidal rocks. *Range*: North Island. *Colour*: Dull grey. *Occurrence*: Fairly common. *Height*: 15-17mm. *Width*: 18-19mm. *REMARKS*: A quite small shell with a sculpture consisting of strong spiral nodulous ridges. Aperture highly callused. The South Island *Diloma* is *D. bicanaliculata lenior*.

18. Blue Top Shell
Diloma nigerrima
(Family Trochidae)

Environment: On decaying kelp at high tide on exposed coasts. *Range*: New Zealand. *Colour*: Exterior: Navy blue with white aperture. Occasionally reddish-purple. Interior: Iridescent in pale pinks and green. *Occurrence*: Common. *Height*: 15-26mm. *Width*: 17-24mm. *REMARKS*: Has a sharp outer lip. The word "*nigerrima*" means "very black".

19. Mudflat Top Shell/ Whētiko
Diloma subrostrata
(Family Trochidae)

Environment: On mud flats. *Range*: New Zealand. *Colour*: Variable: yellowish-grey to purplish-grey. *Occurrence*: Very common. *Height*: 15-31.5mm. *Width*: 16.5-29mm. *REMARKS*: As the measurements show, a shell that varies greatly in sizes. The word "subrostrata" means "slightly beaked" or "slightly curved".

20. Opal Top Shell/ Matangongore
Cantharidus opalus
(Family Trochidae)
Other Māori name: Matangongo

Environment: On seaweed at extreme low tide. *Range*: New Zealand. *Colour*: From orange at top to mauve below. Aperture is iridescent in hues of reds and greens. *Occurrence*: Uncommon. *Height*: 43.5mm. *Width*: 28mm. *REMARKS*: Because of their beautiful nacreous interiors, these shells make fine necklaces and bracelets. The Māori call such strings of small shells "toitoi". The word "*cantharidus*" refers to "an iridescent beetle"; "*opalus*" "an opal".

21. Pink Top Shell/Tihi
Cantharidus purpureus
(Family Trochidae)

Environment: Amongst seaweed and rocks at and below low tide. *Range*: North Island and northern part of South Island. *Colour*: Variable – greenish-grey to rose-pink. *Occurrence*: Fairly common. *Height*: 22-32mm. *Width*: 16-20mm. *REMARKS*: Often only the top of the tall conical spire is pink. It is usually heavily encrusted, "*purpureus*" means "of a purple colour".

22. Tiger Shell/Maurea
Calliostoma tigris
(Family Trochidae)
Other Māori names: Matangongore, Rehoreho, Reoreo, Rereho

Environment: In rock crevices at extreme low tide and deeper water. *Range*: New Zealand. *Colour*: Broad red-brown markings on a white background. *Occurrence*: Rare. *Height*: 45-77.5mm. *Width*: 45-76.5mm. *REMARKS*: A most attractive shell, much sought-after by collectors. Shiny, with delicate whorls over the whole outer surface. Spire very pointed.

23. Pale Tiger Shell/Reoreo
Calliostoma selectum
(Family Trochidae)

Environment: In sand below low tide. *Range*: North Island and northern part of South Island. *Colour*: Pale brown, almost white, with minute speckles of darker brown. *Occurrence*: Not common. *Height*: 37-57mm. *Width*: 45-62mm. *REMARKS*: Another species is occasionally found with *C. selectum* – *C. waikanae*, which is very similar, but is marked with brown blotches.

24. Spotted Tiger Shell
Calliostoma punctulatum
(Family Trochidae)

Environment: On rocks at extreme low tide and in deep water. ***Range***: New Zealand. ***Colour***: Variable - Pinkish-brown to reddish-brown, with many tiny white granulations. ***Occurrence***: Fairly common. ***Height***: (Coarse sculptured) 31.5-35mm. ***Width***: 30-34.5mm. ***REMARKS***: This shell has a solid outer lip. The Stewart Island sub-species, *C. p. stewartiana*, has finer granulations.

25. Wheel Shell/
Kota *Zethalia zelandica*
(Family Umboniinae)

Environment: Deep water. ***Range***: New Zealand. ***Colour***: Exterior: Creamy-white or pinkish with wavy markings of coppery-brown. Occasionally a shell is entirely pink. Interior: Nacreous. ***Occurrence***: Very common. ***Height***: 11.5-13mm. ***Width***: 18.5-22mm. ***REMARKS***: Has a smooth surface with curved raised lines like the spokes of a wheel.

26. Cat's Eye/
Akanakana
Turbo smaragdus
(Family Turbinidae)
Other Māori names: -
A t a a t a , K a i t a n g a t a , Kōrama, Kōramu, Mātangata, Pūpū, Pūpū atamarama, Pūpū Kōrama, Pūpū Kōramu

Environment: On intertidal rocks. ***Range***: New Zealand. ***Colour***: Greenish-black, aperture silvery-white. ***Occurrence***: Common. ***Height***: 40-70mm. ***Width***: 50-73mm. ***REMARKS***: This shell gets its name from the handsome green and white operculum, or lid, with which the animal closes the shell aperture. The animal is herbivorous. The demi-God Tāwhaki, in a story from Māori legend, reputedly placed Cat's Eye shells on his eyes to fool an old witch that he was asleep. He crept out in the early morning to harvest her birds while she herself slept. Hence the saying *He kanohi o Tāwhaki*, "the eyes of Tāwhaki".

At another time Tāwhaki is reputed to have climbed into the heavens, to the tenth level of the skies, and there met a blind woman by the name of Whaitiri. By prayer and by pressing clay moistened with his own spittle to her eyes, he restored her sight. In latter days the Māori invoked Whaitiri's name in prayers to heal eye conditions.

27. Granose Turban/Takawiri
Modelia granosa
(Family Turbinidae)
Other Māori name: Pūpū

Environment: On low tidal rocks and in deep water. *Range*: New Zealand. *Colour*: Light red-brown, tinged with pink and purple. *Occurrence*: Uncommon. *Height*: 40-68mm. *Width*:50-71mm.
REMARKS: The operculum is similar to that of the Cat's Eye but it lacks the green colouring. Southern specimens grow much larger than northern ones. The word "*granosa*", or "granulated", refers to the well-defined rows of granules on the exterior of the shell.

28. Circular Saw/ Ripo Matamata
Astraea heliotropium
(Family Turbinidae)

Environment: Deep water. *Range*: New Zealand. *Colour*: Grey, tinged with pink. *Occurrence*: Uncommon. *Height*: 50-60mm. *Width*: 100-120mm. *REMARKS*: A very large shell keenly sought by collectors. The base is decorated with spiral rows of beading; the whorls are also spirally beaded. The saw-like triangular 'teeth' around the perimeter are much longer on younger shells, than those of adults.

29. Cook's Turban/ Kāeo *Cookia sulcata*
(Family Turbinidae)
Other Māori names: -
Kākara, Karaka, Karekawa, Karikawa, Ngāeo, Ngāruru, Pūpū kaiwhiri, Pūpū karekawa, Rerekākara, Toitoi

Environment: On low tidal rocks. *Range*: New Zealand. *Colour*: Rough grey-brown corrugated coating with aperture pearly white. *Occurrence*: Common. *Height*: 65-80mm. *Width*: 65-90mm.
REMARKS: Usually very encrusted. Has a handsome appearance when cleaned. In the days of cannibalism pins were made from the lower part of the human arm in order to denigrate the original owner of the arm and his family. These pins were used to pry loose the animals from Cook's Turban and other shells. The saying *te ngāruru e piri te toka* refers to the attribute of this shellfish in clinging to its rocky home, a saying used of a steadfast person, one who does not yield.

30. Black Nerita/Matangārahu
Nerita atramentosa melanotragus
(Family Neritidae)
Other Māori names: Matapura, Ngārahu tatawa, Ngārahu taua, Peke

Environment: On intertidal rocks. *Range*: North Island. *Colour*: Blackish. Aperture is white. The operculum is pink and grey. *Occurrence*: Common. *Height*: 22-30mm. *Width*: 25-32mm. *REMARKS*: Esteemed as food. Sub-species of the Australian *N. atramentosa* it is the sole member of this tropical family in New Zealand. It keeps to the warmer waters of the North Island.

32. Horn Shell/Koeti
Zeacumantus lutulentus
(Family Batillariidae)

Environment: On mud flats. *Range*: North Island and northern part of South Island. *Colour*: Greyish-brownish. Aperture yellow to dark brown. *Occurrence*: Very common. *Height*: 24-29mm. *Width*: 8.5-10mm. *REMARKS*: The word "*lutulentus*" means "muddy". In mangrove swamps in dry seasons the animal closes its aperture with its operculum and hangs, suspended by glutinous threads, to a small branch or root. The small black *Z. subcarinatus* is common in rock pools.

31. Periwinkle/Mākerekere
Nodilittorina cincta
(Family Littorinidae)

Environment: On high tidal rocks. *Range*: New Zealand. *Colour*: Grey. *Occurrence*: Very common. *Height*: 13.5-19.5mm. *Width*: 9-11mm. *REMARKS*: The patterns and spirals seen on shells of the periwinkle family are said to have provided design inspiration for the Māori tattooer. Like Cook's turban, this shellfish was taken for food by the Māori in earlier times, using pins made from the lower human arm to extract the animal.

33. Turret Shell/Papatai
Maoricolpus roseus
(Family Turritellidae)

Environment: Low tide, in sandy, muddy or rocky areas, and in deeper water, almost buried in the sand, point down. *Range*: New Zealand. *Colour*: Yellowish, brownish, purplish. *Occurrence*: Common. *Height*: 53-86.5mm. *Width*: 19-25mm. *REMARKS*: A rather large shell with variable sculpture, polished inside and out, and with a thin outer lip. The slender sub-species *M. roseus manukauensis* is found only in the Manukau Harbour

34. Pagoda Turret Shell/ Karire
Zeacolpus pagoda
(Family Turritellidae)

Environment: In sand, below low tide. *Range*: Northland. *Colour*: Pinkish-white with brown markings. *Occurrence*: Fairly common. *Height*: 20-37mm. *Width*: 6.2-9mm. *REMARKS*: A thin and fragile shell. Its first two whorls are smooth, the next two have raised middle ridges that are situated lower down on later whorls.

35. Stewart Island Turret Shell
Zeacolpus symmetricus
(Family Turritellidae)

Environment: In sand, below low tide. *Range*: Southern part of the South Island and Stewart Island. *Colour*: Dull white to pale buff. *Occurrence*: Common. *Height*: 17mm. *Width*: 6mm. *REMARKS*: A small, rather thin and translucent shell with about fourteen whorls.

36. Wentletrap/Tōtoro
Cirsotrema zelebori
(Family Epitoniidae)

Environment: In sandy areas, below low tide. *Range*: New Zealand. *Colour*: White. Operculum black. *Occurrence*: Common. *Height*: 24-31mm. *Width*: 10-11mm. *REMARKS*: A very dainty shell that is sometimes known as "The Curly", the numerous ribs across the ten or eleven whorls being very prominent. More common in the north than the south. Named "*zelebori*" after the naturalist Zelebor.

37. Violet Snail/Kararua
Janthina janthina
(Family Janthinidae)

Environment: Pelagic, inhabiting open oceans. *Range*: New Zealand – World. *Colour*: Pale translucent violet. *Occurrence*: Common at times, on open beaches. *Height*: 21-38mm. *Width*: 26-40mm. *REMARKS*: Swept ashore during gales. The animal, when touched, emits a violet-coloured fluid, the same colour as the shell. This fluid was used in ancient times in Europe to produce Tyrian purple dye. *J. exigua* is also common at times on open beaches. It is smaller with deeper violet colouring and a pointed spire. Both species are very fragile.

38. Circular Slipper Shell/ Ngākihi kopia
Sigapatella novaezelandiae
(Family Calyptraeidae)

Environment: On stones or shells at low tide. ***Range***: New Zealand. ***Colour***: Whitish tinged with purple under a brown epidermis. ***Occurrence***: Common. ***Height***: 7-25mm. ***Width***: 18-33mm.
REMARKS: The height of the shell is very variable. Young shells are mostly white at the base with purplish-brown roofs.

40. Ribbed Slipper Shell/ Ngākihi hiwihiwi
Crepidula costata
(Family Calyptraeidae)

Environment: On shells and rocks at low tide. ***Range***: North Island. ***Colour***: Exterior: Light brown, rough. Interior: White with brown rays. ***Occurrence***: Common. ***Length***: 34-62mm. ***Width***: 24-43mm.
REMARKS: The outer shell is strongly ribbed, hence the description "*costata*" or "ribbed". The interior is smooth and porcelain-like. The clear separation line between the white section and that of the brown rays, and the shape of the shell, giving rise to the "slipper shell" name.

39. White Slipper Shell/ Ngākihi tea
Crepidula monoxyla
(Family Calyptraeidae)

Environment: On rocks, and in the apertures of dead univalves; low tide to deeper water. ***Range***: New Zealand. ***Colour***: White or yellowish-white. ***Occurrence***: Common. ***Length***: 23.5-34mm. ***Width***: 15-29.5mm. ***REMARKS***: Often found on Cat's Eyes shells, *Turbo smaragdus*. *C. monoxyla* is very variable in shape, those found on rocks being humped (convex), while those found in the apertures of dead univalves are quite flat, or concave above.

41. Carrier Shell/ Papa kotakota
Xenophora neozelanica
(Family Xenophoridae)

Environment: Deep water. ***Range***: North Island. ***Colour***: White with brown on raised ridges. ***Occurrence***: Uncommon. ***Height***: 42-45mm. ***Width***: 60-62mm.
REMARKS: This strange mollusc camouflages itself by cementing dead shells, small stones and other debris to its shell. Bivalves cemented have the inner side of the valve uppermost. Formerly called *X. corrugata*, "*corrugata*" is "wrinkled".

42. Small Ostrich Foot/
Takai *Pelicaria vermis*
(Family Struthiolariidae)
Other Māori name: Tukai

Environment: In sand at and below low tide. *Range*: North Island and northern part of South Island. *Colour*: Off-white to red-brown, with yellow lip. *Occurrence*: Fairly common. *Height*: 32-54mm. *Width*: 20.5-35mm. *REMARKS*: The mouth of this species is said to resemble the shape of an ostrich's foot, hence the name. The name "*vermis*" or "worm" refers to the suture, or line of union between one whorl and the next. This line is said to be like the track of a worm

43. Ostrich Foot/
Totorere
Struthiolaria papulosa
(Family Struthiolariidae)
Other Māori Names:
Kaikaikaroro, Takai

Environment: Buried in sand or mud, at and below low tide. *Range*: New Zealand. *Colour*: Exterior: Light brown with wavy markings. The lip is of white or pale yellow. Interior: Purplish-brown. *Occurrence*: Common. *Height*: 72-80mm. *Width*: 46-48mm. *REMARKS*: Like *S. vermis*, the operculum is very small and claw-shaped. "*Papulosa*" refers to the shell having papules or small pimples. Also called the Ring Shell. The peristome, or ring-like border, of the adult shell is often found loose on the beach in late summer. It contains the embryonic shells of this mollusc. This shell was strung by Māori on flax for necklaces. The whole shell was used, a hole being pierced near the last whorl, near the mouth.

44. New Zealand Cowry
Trivia merces
(Family Triviidae)

Environment: Deep water. Live specimens are very rare. *Range*: Northern part of New Zealand. *Colour*: Pinkish-white with several reddish-brown blotches. *Occurrence*: Rare. *Length*: 13-14mm. *Width*: 9.5mm. *REMARKS*: This shellfish has the power of dissolving or altering its form, its saliva containing muriatic acid. It can dissolve the inner part of its shell or deposit new layers on the outside.

45. Necklace Shell/ Ngaere *Tanea Zelandica*
(Family Naticidae)

Environment: In sand at and below low tide. *Range*: New Zealand. *Colour*: Light brown with six rows of small brown V-shaped dots. *Occurrence*: Fairly common. *Height*: 24.75-33mm. *Width*: 21-31.5mm. *REMARKS*: This mollusc has a shelly white operculum. The female makes a sand collar (also illustrated here) to conceal her eggs. To form this, many fine grains of sand are cemented together.

46. Helmet Shell/ Pūpū māeneene *Semicassis pyrum*
(Family Cassididae)

Environment: In sand at and below low tide. *Range*: North Island. *Colour*: Creamy-pink, with small reddish-brown blotches. *Occurrence*: Fairly common. *Height*: 57-89mm. *Width*: 42-60mm. *REMARKS*: This is the most common of seven species of *Semicassis* found in New Zealand waters. A row of small nodules is to be found around the shoulder. The spire is low and sharply pointed. The operculum is small, yellowish, and shaped like an open fan.

47. Cask Shell/ Pūpū tangimoana *Tonna cerevisina*
(Family Tonnidae)
Other Māori name: Pūpū waitai

Environment: Deep water. *Range*: North Island. *Colour*: Pale creamy-pink with a few brown blotches. Epidermis: Light brown. *Occurrence*: Rare. *Height*: 157-230mm. *Width*: 124-188mm. *REMARKS*: A large and fragile shell with strong broad whorls sometimes washed ashore after gales.

48. Spengler's Trumpet/ Pū moana *Cabestana spengleri*
(Family Cymatiidae)
Other Māori names: Pūpū kākara, Pūtara, Pūtātara

Environment: On rocks at extreme low tide and deeper water. *Range*: New Zealand. *Colour*: Yellowish-brown with darker narrow bands of chestnut brown. Aperture is porcelain white. *Occurrence*: Fairly common. *Height*: 81-160mm. *REMARKS*: Finely ribbed, with shouldered whorls. The outer lip is heavily indented. A carnivorous animal, eating other bivalves. This species also finds a home from South Queensland to Tasmania, as well as South Australia.

49. Hairy Triton
Cymatium parthenopeum
(Family Ranellidae)

Environment: Rocky ground at extreme low tide and in deeper water. ***Range***: North Island and Golden Bay, Nelson. ***Colour***: Grey to brown. Columella is vividly striped in black and white. ***Occurrence***: Uncommon. ***Height***: 86-100mm. ***Width***: 52-59mm.
REMARKS: Somewhat fatter than Spengler's Trumpet. This shell in the live state has a very hairy dark brown epidermis that is washed away by the friction of sand at the seashore when it is cast up.

50. Trumpet Shell/Awanui
Charonia lampas rubicunda
(Family Ranellidae)
Other Māori names:-
Pūhāureroa, Pūwhāureroa, Pūpū tara, Pūpū tātara, Pūtaratara

Environment: Rocky ground at extreme low tide and in deeper water. ***Range***: New Zealand. ***Colour***: White (deep sea), reddish-brown (shallows), marked with various shades of brown. Aperture is white. ***Occurrence***: Uncommon. ***Height***: 143-171mm. ***Width***: 85-108mm. ***REMARKS***: The brown markings are more obvious in juvenile specimens. Adults usually have dull colouring and heavy encrustation. Used by the Māori as a conch shell or horn. It was blown at ceremonies of various kinds and to give warning signals

51. Swollen Triton
Argobuccinum pustulosum tumidum (Family Ranellidae)

Environment: Rocky ground at low tide and in deeper water. ***Range***: New Zealand. ***Colour***: Red-brown, with many narrow dark brown stripes. The aperture is white to lilac-grey. ***Occurrence***: Uncommon, though more common in the south. ***Height***: 57-114mm. ***Width***: 36-79mm.
REMARKS: The epidermis is thin, of light grey-brown colour, and velvety. The sculpture of the shell is fine and closely spiralled.

52. Australian Triton
Ranella australasia
(Family Ranellidae)

Environment: Rocky ground at and below low tide. ***Range***: New Zealand. ***Colour***: Variable. Yellowish-brown to reddish-brown to dark purplish-brown. White aperture and velvety epidermis. ***Occurrence***: Uncommon. More common in the north. ***Height***: 67-94mm. ***Width***: 36-54mm.
REMARKS: Common from New South Wales to southern Western Australia. Like the Trumpet Shell, the top of this shell was cut off and a carved wooden mouthpiece added to make a signalling instrument.

53. Spiny Murex/ Pūpū tarataratea
Poirieria zelandica
(Family Muricidae)

Environment: Shallow to deep water. *Range*: New Zealand. *Colour*: White. *Occurrence*: Uncommon. *Height*: 48-65mm (with spines). *Width*: 35-79mm (without spines). *REMARKS*: This attractive shell, with its long, slender spines, is much sought after by collectors. Molluscs of this species with shorter spines come from shallow waters. Sometimes known as the spider shell because of its spines that look like spiders' legs.

54. Octagonal Murex/ Pūpū taratara
Murexsul octagonus
(Family Muricidae)

Environment: Rocky ground at low tide and in deep water. *Range*: North Island. *Colour*: Light to dark brown. Interior of the aperture is white, often tinted with purple or light brown. *Height*: 26-60mm. *Width*: 13.5-32mm. *REMARKS*: A thick and solid shell. Deepwater specimens are white with rows of short spines, and were previously known as *M. cuvierensis*.

55. Large Trophon
Xymene ambiguus
(Family Muricidae)

Environment: In sand or mud, at and below low tide. *Range*: New Zealand. *Colour*: Yellowish-white. *Occurrence*: Common. *Height*: 35-52mm (male shell). *Width*: 18.5-31mm (male shell). *Height*: 29.5-71mm (female shell). *Width*: 16-33mm (female shell). *REMARKS*: The female animal lays her eggs on the smooth back of the male shell. The eggs are flat white capsules. The male shell is rather smaller than the female's.

56. Common Trophon
Xymene plebeius
(Family Muricidae)

Environment: On mudflats. *Range*: New Zealand. *Colour*: Generally creamy-buff in colour with purplish-brown oval apertures. *Occurrence*: Common. *Height*: 15-22.5mm. *Width*: 7.5-11.5mm.
REMARKS: A small carnivorous snail often found in harbour waters. It bores into other shellfish for food.

57. Rock Trophon
Paratrophon patens
(Family Muricidae)

Environment: On low tidal rocks. *Range*: New Zealand except Northland. *Colour*: Creamy-white. Aperture: Brown. *Occurrence*: Fairly common. **Height**: 21-24mm. **Width**: 14.25-15.75mm. **REMARKS**: The sculpture of this shell is most variable. Some specimens have strong spiral ribs, others only traces of them. A similar, but smaller species, *P. cheesemani cheesemani*, is found on the west coast of Northland.

58. Stanger's Trophon/
Paratrophon quoyi
(Family Muricidae)

Environment: On intertidal rocks. *Range*: Northland and Bay of Plenty. *Colour*: Buff with white lip and purplish aperture. *Occurrence*: Fairly common. **Height**: 24-35mm. **Width**: 13.5-17.5mm. **REMARKS**: This shell is of rugged sculpture with six or seven strong whorls and a flattish base.

59. White Rock Shell/
Hopetea *Dicathais orbita*
(Family Muricidae)
Other Māori name: Tāwiri

Environment: In rock crevices at low tide. *Range*: North Island and northern part of South Island. *Colour*: Dull creamy-white. Reddish-brown on open coasts. Aperture tinged with yellow. *Occurrence*: Common. **Height**: 58-118mm. **Width**: 40-73mm. **REMARKS**: A large heavily sculptured shell with rounded spiral ridges on the whorls. Sculpture smoother when from sheltered areas.

60. Dark Rock Shell/
Kāeo *Haustrum haustorium*
(Family Muricidae)
Other Māori names: Kākara, Ngāeo

Environment: On intertidal rocks on sheltered coasts. *Range*: New Zealand. *Colour*: Dull purple with fine lines of light purple. Aperture is whitish with an inner dark patch. *Occurrence*: Common. **Height**: 54-65mm. **Width**: 35-45mm. **REMARKS**: Another shell first brought to Europe by Captain Cook's 18th Century expeditions. Has short spire and large ear-shaped aperture. Sometimes called the "dog winkle". A carnivorous animal feeding on top shells.

61. Oyster Borer
Kaikai tio
Lepsiella scobina
(Family Muricidae)

Environment: On intertidal rocks. ***Range***: New Zealand. ***Colour***: Buff with spiral lines in brown. Aperture – dark brown. ***Occurrence***: Common. ***Height***: 17-21.5mm. ***Width***: 12-14mm. ***REMARKS***: This shell takes its common name from its depredations of oysters beds, but it bores into mussels and barnacles with equal vigour. Smallest of the New Zealand shell borers, but very numerous.

62. Many-Lined Whelk/ Huamutu
Buccinulum linea linea
(Family Buccinidae)

Environment: Low tidal rocks. ***Range***: North Island and northern part of South Island. ***Colour***: Variable. Mostly pale brown with many fine purple lines. ***Occurrence***: Fairly common. ***Height***: 35-49mm. ***Width***: 17-23mm. ***REMARKS***: There are marked variations between the thirteen shells of the *Buccinulum* species in the shapes of their spires and the spaces between their spiral lines.

63. Siphon Whelk/
Kākara *Penion sulcatus*
(Family Buccinidae)
Other Māori names: - Huamutu, Pūpū kōihi

Environment: Rocky areas. ***Range***: North Island. ***Colour***: Grey to white, with white aperture. ***Occurrence***: Fairly common. ***Height***: 127-162mm. ***Width***: 62-66mm. ***REMARKS***: A large shell with strong shoulders and strong whorls alternating with narrower ones. Beach worn specimens are distinctively marked with dark-brown and white lines.

64. Deep Water Siphon Whelk/ Kākara nui
Penion cuvierianus
(Family Buccinidae)
Other Māori name: Huamutu

Environment: Deep water. ***Range***: North Island. ***Colour***: Buff with sparse brown markings. ***Occurrence***: Rare. ***Height***: 140-235mm. ***Width***: 57-113mm. ***REMARKS***: Another very large shell with varying sculpture, shoulder angle and whorl shape. A close relative is *Penion maxima* of Australia.

65. Southern Siphon Whelk/ Huamutu
Penion mandarina
(Family Buccinidae)

Environment: Low tide and deeper water, rocky ground. ***Range***: Northern part of South Island. ***Colour***: Greyish to brownish. ***Occurrence***: Uncommon. ***Height***: 126-144mm. ***Width***: 56-62mm.
REMARKS: Its sculpture consists of equidistant spirals, with one or two finer cords, and some spiral threads on the lower part.

66. Knobbed Whelk/ Kākara *Austrofusus glans*
(Family Buccinidae)

Environment: In sand, from low tide to deep water. ***Range***: New Zealand. ***Colour***: Yellowish-white with pinkish-brown markings. A dark spiral band is often to be noted encircling each whorl. White lip. ***Occurrence***: Common. ***Height***: 43-74mm. ***Width***: 26-43mm. ***REMARKS***: Shallow water specimens have a paler brown coloured periostracum than those from deep water.

67. Speckled Whelk/ Kāwari *Cominella adspersa*
(Family Buccinidae)

Environment: On mudflats and in deeper water. ***Range***: North Island and northern part of South Island. ***Colour***: Variable. Buff, liberally speckled with grey, to chocolate-brown. ***Occurrence***: Very common. ***Height***: 34-70mm. ***Width***: 23-43mm. ***REMARKS***: The plump northern form previously known as *C. adspersa melo* is now considered to be the same as *C. adspersa*. The egg capsules are whitish and have the appearance of plastic. One of the illustrated specimens here illustrated is covered in these capsules.

68. Spotted Whelk/Kāwari
Cominella maculosa
(Family Buccinidae)

Environment: On mudflats and rock platforms. ***Range***: North Island and northern part of South Island. ***Colour***: Mostly chocolate-brown but also greenish-yellow with many reddish-brown spots. Sometimes unspotted specimens occur. Aperture dark brown, outer lip brown. ***Occurrence***: Very common. ***Height***: 38-46.5mm. ***Width***: 21-24mm. ***REMARKS***: Difficult to get a good specimen as the spire, or upper part of the shell, is frequently worm-eaten.

69. Quoy's Whelk
Cominella quoyana
(Family Buccinidae)

Environment: In sand, low tide to deeper water. ***Range***: North Island. ***Colour***: Buff with fine spiral lines and square dots of darker colour. ***Occurrence***: Fairly common. ***Height***: 20-21.5mm. ***Width***: 8.5-11mm. ***REMARKS***: Like other *Cominella*, a carnivorous scavenger of its territory. Named after the naturalist Quoy who collected shells during the visit to New Zealand of Dumont D'Urville in the 1820's and 1830's.

70. Mud Whelk
Cominella glandiformis
(Family Buccinidae)

Environment: On mudflats. ***Range***: New Zealand. ***Colour***: Variable - whitish, also shades of grey, and rarely orange. Bands of purplish-brown on spire, aperture also purplish-brown. ***Occurrence***: Very common. ***Height***: 22-24.5mm. ***Width***: 12-13mm. ***REMARKS***: The name *glandiformis* means "acorn-shaped", not perhaps a very accurate description. A scavenger of dead or dying bivalves on cockle and pipi beds.

71. Red-Mouthed Whelk
Cominella virgata
(Family Buccinidae)

Environment: Rocky areas, intertidal. ***Range***: North Island and northern part of South Island. ***Colour***: Greyish with fine dark brown lines. The aperture is reddish-brown with reddish-orange lip and columella. ***Occurrence***: Common. ***Height***: 30-42mm. ***Width***: 15-20mm. ***REMARKS***: The subspecies *C. virgata brookesi* is found in Parengarenga Harbour, Northland. It has well-developed nodules, whereas *C. virgata* is almost completely smooth.

72. Stewart Island Whelk
Cominella nassoides
(Family Buccinidae)

Environment: In sand at low tide and deeper water. ***Range***: Foveaux Strait and Stewart Island. ***Colour***: Exterior: Beige, light orange or grey. Interior is brownish. ***Occurrence***: Fairly common. ***Height***: 30-56mm. ***Width***: 15.75-24mm. ***REMARKS***: The deepwater form, previously known as *C. fouveauxana* is now considered to be the same as *C. nassoides*. Sculpture consists of strong, well-spaced rounded ribs with spiral lines intersecting.

73. Pagoda Shell
Coluzea spiralis
(Family Turbinellidae)

Environment: Amongst sand and rubble in deep water. *Range*: North Island and northern part of South Island. *Colour*: Very pale brown with even lighter brown markings. *Occurrence*: Very rare. *Height*: 44.7-107.2mm. *Width*: 15.1mm - 28.3mm. *REMARKS*: A very long and striking-looking shell keenly sought by collectors.

74. Brown Olive
Amalda mucronata
(Family Olividae)

Environment: Below low tide in sand. *Range*: North Island and northern part of South Island. *Colour*: Spire is pale orange-brown, fawn in the middle, to brown at end. Has two white bands. *Occurrence*: Not common. *Height*: 30-61mm. *Width*: 13.5-27mm. *REMARKS*: The shell has a large aperture, so large in fact that the living animal when fully extended completely covers its shell.

75. Southern Olive/Pūpū pīataata *Amalda australis*
(Family Olividae) Other Māori names: Pūpū rore, Tikoaka, Uere

Environment: In sand, at and below low tide. *Range*: North Island and northern part of South Island. *Colour*: Exterior: Shiny light blue-grey centre with brown at each end, and a little white. Interior: Purplish. *Occurrence*: Common. *Height*: 32-40mm. *Width*: 16-19mm. *REMARKS*: The Māori used the shells in necklaces and in post-colonial days selected the medium-sized shells as buttons for clothing.

76. Arabic Volute/Pūhaureroa *Alcithoe arabica*
(Family Volutidae) Other Māori names: Pūwhaureroa, Pūpū rore, Tākupu, Uere, Whakai-a-tama

Environment: Below or at low tide in sand or mud. *Range*: North Island and northern part of South Island. *Colour*: Base of pale red buff with purplish-brown zigzag markings. *Occurrence*: Fairly common. *Height*: 116-126mm. *Width*: 42.5-51.5mm. *REMARKS*: This shell has nodules around the shoulders. Like other species of *Alcithoe*, *A. arabica* lays a round, shelly white egg capsule from which 2 to 4 baby animals hatch out. The shell was used by the Māori as a conch or horn blown to give signals etc. A wooden mouthpiece was attached by fibre lashings.

77. Southern Volute/ Whakai-a-tama
Alcithoe swainsoni
(Family Volutidae)

Environment: In sand from low tide to deep water. *Range*: New Zealand, except Auckland. *Colour*: Buff, with reddish-brown zigzag markings. *Occurrence*: Fairly common. *Height*: 98-201mm. *Width*: 32-41mm. *REMARKS*: Considered now as a relative of *A. arabica*.

79. Calva Volute
Alcithoe calva
(Family Volutidae)

Environment: Deep water. *Range*: Cook Strait southwards. *Colour*: Pinkish-buff with inconspicuous zigzag markings. *Occurrence*: Rare. *Height*: 151-177mm. *Width*: 53-64mm. *REMARKS*: A similar deepwater northern species is *A. jaculoides*, which has nodules, whilst *A. calva* is quite smooth.

78. Depressed Volute/ Pūpū rore
Alcithoe arabica depressa
(Family Volutidae)

Environment: Below low tide, in sand. *Range*: Northland. *Colour*: Buff with many dark zigzag markings. *Occurrence*: Uncommon. *Height*: 70-80mm. *Width*: 41-88mm. *REMARKS*: Also considered as another form of *A. arabica*. It has a smaller, squatter shell, with a lower spire.

80. Little Volute
Alcithoe fusus
(Family Volutidae)

Environment: Below low tide, in sand. *Range*: South Island and lower part of North Island. *Colour*: Buff, with wavy reddish-brown markings. *Occurrence*: Rare. *Height*: 46-73mm. *Width*: 17-29mm. *REMARKS*: This species is replaced from Bay of Plenty northwards by the subspecies *A. haurakiensis* which has small nodules, whilst *A. fusus* is smooth.

81. Golden Volute
Provocator mirabilis
(Family Volutidae)

Environment: Very deep water. *Range*: New Zealand, except Northland. *Colour*: Shining salmon orange. *Occurrence*: Very rare. **Height**: 112-140mm. **Width**: 39-48mm. **REMARKS**: This shell is either tall-spired, slender and of light build, or lower-spired and rather solid. A choice collector's item, as rare as it is beautiful.

82. Pink Tower Shell/
Torire *Phenatoma rosea*
(Family Conidae)

Environment: Below low tide in sand. *Range*: New Zealand. *Colour*: Pink. *Occurrence*: Not common. **Height**: 29-34mm. **Width**: 11-11.25mm. **REMARKS**. This shell is the best known of the more than one hundred species of *Conidae* found in New Zealand waters. Easily recognised from the anal siphonal notch in the outer lip.

83. New Zealand Auger
Pervicacia tristis
(Family Terebridae)

Environment: In sand below low tide. *Range*: North Island and northern part of South Island. *Colour*: Chocolate with a yellow or white spiral band. *Occurrence*: Fairly common. **Height**: 13.5-22mm. **Width**: 4.5-7.5mm. **REMARKS**: This is the small size New Zealand representative of a family that has many large and very colourful tropical species.

84. Brown Bubble Shell/
Pūpū waharoa *Bulla quoyii*
(Family Bullidae)

Environment: Mudflats at and below low tide. *Range*: North Island. *Colour*: Yellowish-brown with brown shades. Sometimes marbled with purplish-grey, or with white dots. *Occurrence*: Fairly common. **Height**: 33.5-48mm. **Width**: 21-29mm. **REMARKS**: This is a quite large shell named for Quoy, the French naturalist, who visited New Zealand in the early 19th century. *Bulla* here means "bubble", referring to the almost-round shape of the shell.

85. White Bubble Shell/ Pūpū tuatea
Haminoea zelandiae
(Family Haminoeidae)

Environment: Mudflats at low tide. ***Range***: North Island and northern part of South Island. ***Colour***: Pale yellowish-brown or plain white when the epidermis has been lost. ***Occurrence***: Fairly common. ***Height***: 17-27mm. ***Width***: 13-22mm. ***REMARKS***: A thin and fragile shell with a smooth surface, partially internal in a grey slug, hence sometimes called the sea snail.

86. Mud Snail/Karahū
Amphibola crenata
(Family Amphibolidae)
Other Māori names: - Karahue, Koriakai, Tītiko, Wētiwha, Whētiko, Whētikotiko.

Environment: On mudflats. ***Range***: New Zealand. ***Colour***: Yellowish-brown. The columella is mauve-brown. ***Occurrence***: Very common. ***Height***: 21.5-30.5mm. ***Width***: 22.5-29.5mm. ***REMARKS***: About the same size as a garden snail, the animal inside spawning an egg mass of between five and ten thousand eggs. The shell's animal feeds on vegetable matter contained in mudflat mud. A morsel much enjoyed by both Māori and Pākehā.

87. Small Siphon Limpet/ Ngākihi awaawa
Siphonaria australis
(Family Siphonariidae)

Environment: On high tidal rocks. ***Range***: New Zealand. ***Colour***: Exterior: Brown to chestnut coloured. Interior: Liver-brown but darker brown over the central area, polished and horseshoe shaped. ***Occurrence***: Common. ***Length***: 20-27mm. ***Width***: 15-22mm. ***REMARKS***: Both siphon limpets described here have a siphonal groove that allows them to inhale and exhale water. They browse on algae.

88. Large Siphon Limpet
Benhamina obliquata
(Family Siphonariidae)

Environment: On high tidal rocks on exposed coats. ***Range***: South Island. ***Colour***: Exterior: Light to dark brown, usually encrusted. Interior: Beige to orange with brown and white margins. ***Occurrence***: Common but local. ***Length***: 34-65mm. ***Width***: 23-44mm. ***REMARKS***: The largest of the *Siphonariidae* family it has 20 or 20 ribs radiating from the apex of the shell. The siphonal groove is shallow and very broad.

TUSK SHELL

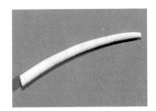

89. Tusk Shell/
Hangaroa *Antalis nana*
(Family Dentaliidae)
Other Māori names:-
Kōmore, Pipi kōmore, Pipi taiari

Environment: Really a type of limpet that has chosen mud or sand in deep water as its habitat. *Range*: New Zealand. *Colour*: White. *Occurrence*: Uncommon. *Length*: 20.7-37.7mm. *Width*: 2-3mm. *REMARKS*: A tube-shaped shell open at both ends. The animal has a foot used for burrowing and a many-tentacled head. Employed in former times by the Māori to make necklaces comprising six to nine strands. A compliment paid to a woman with sparkling white teeth is *Me he pipi taiari*: "Your teeth are like the pipi taiari".

BIVALVES

90. Razor Mussel/
Kuku para
Solemya parkinsonii
(Family Solemyidae)
Other Māori name: Kute

Environment: Deeply buried in silty mud. *Range*: New Zealand. *Colour*: Exterior: Covered by a shining chestnut-brown epidermis with fringed edge. Interior: Pale grey. *Occurrence*: Fairly common. *Length*: 38-55mm. *Height*: 14-20mm. *REMARKS*: A fragile shell. When the mantle is protruding and the animal moving through shallow water by rhythmic extensions of its foot, the shellfish looks just like a pink and purple flower in bloom.

91. Mallet Shell
Neilo australis
(Family Malletiidae)

Environment: In black mud in deep water. *Range*: New Zealand. *Colour*: Exterior: Grey-brown. Interior: White, nacreous. *Occurrence*: Rare. *Length*: 34-42.6mm. *Height*: 17-21.1mm. *REMARKS*: An elongated, rather dull-surfaced shell with well-spaced lines of darker colours. The hinge consists of a long row of fine teeth, about 60 in number.

92. Ark Shell/Tūroro
Barbatia novaezealandiae
(Family Arcidae)

Environment: Under stones at low tide and in deeper water. ***Range***: New Zealand. ***Colour***: Exterior: Dull white to brown with dark brown hairy epidermis. Interior: White, tinged with brownish streaks. ***Occurrence***: Fairly common. ***Length***: 45-74mm. ***Height***: 24-42mm. ***REMARKS***: The animal fastens itself to the undersides of rocks by a strong fibre, or byssus

93. Small Dog Cockle/Kuakua
Glycymeris modesta
(Family Glycymerididae)

Environment: In sand, low tide to deeper water. ***Range***: New Zealand. ***Colour***: Exterior: Ranges from white to reddish-brown with striking zigzag or radial patterns in different shades of reddish-brown. Interior: Porcellanous, whitish and brown. ***Occurrence***: Common. ***Height***: 18.5-25mm. ***Width***: 20-28mm. ***REMARKS***: A small shell with very fine sculpture.

94. Large Dog Cockle/Kuhakuha
Tucetona laticostata
(Family Glycymerididae)

Environment: In sand below low tide. ***Range***: New Zealand. ***Colour***: Exterior: Adults – Dull yellowish to reddish-brown. Interior: White or brownish, porcellanous. Exterior: Juveniles – Prettily marked in deep pink on a white background. ***Occurrence***: Common. ***Height***: 52-118mm. ***Width***: 52-103mm. ***REMARKS***: Distinct growth rings and radiating ridges. The ridges of younger shells wear off with age. Often cast up on ocean beaches.

THE MUSSEL FAMILY

Māori mythology informs us that mussels are the offspring of Kaukau and Te Ropuwai, and were placed under the protection of Wharerimu and Wharepapa, personifications of seaweed and rocks. This may have come about because of a fight that took place between the mussel and cockle families in distant times. When the mussel family thrust out their "tongues" they got them clogged up with sand and so they were defeated. They were forced to leave the sandy habitat of the cockles and take up their abode among rocks and seaweed instead. The mussel family remains in these surroundings to this day.

Saltwater mussels supplied an important source of food in earlier times. A rock where mussels were to be found in abundance was called a "pātiotio". "A rock covered with mussels" (*He toka kukupara*) is a phrase to describe where lots of mussels are to be found. Mussels were taken with "koikoi" from the shallows, and with "pūrau" (pointed sticks) from deep water. They were cooked, and if desired as a reserve food, threaded onto strips of flax, and dried in the sun. When out travelling through the country one could always rely on finding a supply of shellfish like mussels as sustenance. Simply place the mussels in a heap, pile dry fern on top, and set alight.

One use of mussel shells was to protect kūmara (sweet potato) plantations from predatory rats. Long lines were threaded with broken pieces of mussel shells and laid across the beds. During the night the ends of the lines were jerked by the old men of the village in order to frighten the rats away. The shells also came in handy as tweezers to trim a beard, as scrapers for flax and kūmara, etc.

Proverbs attest to the potency of the mussel. *Me te tumu kuku*, "like kuku on a mussel bed" is a reference to a party of people who persist in remaining indoors when asked to come outside. Their chief exercises great authority. *Me te kuku ka kopi*, "like the neat closing of a shell", refers to the lips of a severed enemy head. If the lips had been sewn together neatly, this was the comment. That is, well done!

95. Blue Mussel/Kuku
Mytilus galloprovincialis
(Family Mytilidae)
Other Māori names: -
Kūtai, Pipi, Pōrohe, Toretore, Torewai, Toritori

Environment: On rocks at low tide. *Range:* New Zealand, but uncommon in the north. *Colour:* Exterior: Blue to black with a little white towards the umbones. Interior: Bluish to violet-black. *Occurrence*: Common. *Anterior – posterior:* 61-118mm. *Dorso – ventral:* 31-63mm. *REMARKS:* Its sculpture consists of fine concentric growth lines, its narrow hinge has several teeth. Its byssus (fibre threads) attach the shellfish to flat surfaces.

96. Green Mussel/Kuku
Perna canaliculus
(Family Mytilidae)
Other Māori names: -
Kukutai, Kūtai, Pōrohe, Toretore, Torewai, Toritori

Environment: On rocks, low tide to deeper water. *Range:* New Zealand. *Colour:* Exterior: Juveniles – Sometimes red to light yellow, or bright green. Adults – Brown to black, edged with green. Interior: Somewhat iridescent purple. *Occurrence:* Very common. *Anterior – posterior:* 96-163mm. *Dorso – ventral:* 47-94mm. *REMARKS:* The largest New Zealand mussel, commoner in the north than the south. It is widely farmed both for its food and for its medicinal values.

97. Ribbed Mussel/Kuku
Aulacomya atra maoriana
(Family Mytilidae)
Other Māori name: Pūkanikani

Environment: On low tidal rocks. *Range:* Rare in the North Island but common in the South. *Colour:* Exterior: Whitish to dull purple. Interior: Shiny reddish-purple. *Occurrence:* Common. *Anterior – posterior:* 56-84mm. *Dorso – ventral* : 26-37mm. *REMARKS:* The sculpture consists of thick raised undulating ribs radiating from the beak that reinforce the shell and make it very strong. The only mussel species to be found among bull kelp where the wave action is often severe.

98. Nesting Mussel/Korona
Modiolarca impacta
(Family Mytilidae)
Other Māori names: -
Kuku mau toka, Kuku moe toka, Kukupara, Kuku weu, Torewai

Environment: On rocks at low tide and in deeper water. *Range:* New Zealand. *Colour:* Exterior: Variable, from yellow to leathery deep olive green. Interior: Iridescent pink to purple. *Occurrence:* Common. *Anterior – posterior:* 33.5-46mm. *Dorso – ventral* : 24-28mm. *REMARKS:* The nesting mussel lives concealed in the "nests" of fibrous brown byssus threads that it builds. Several dozen individuals of all sizes may live in the one nest.

99. Date Mussel/Taitaki
Zelithophaga truncata
(Family Mytilidae)

Environment: Boring into soft rock at low tide and deeper water. *Range:* New Zealand.

Colour: Exterior: Dark brown. Interior: Bluish-white to purple. *Occurrence*: Not common. *Length*: 40-49mm. *Height*: 13.5-15.5mm. *REMARKS*: The common name given the date mussel comes from its resemblance to a date stone, both in shape and in colour. It lives in its hole in the rocks, with only a tiny mark to show where it is hidden. It attaches itself at the middle of its body with byssus threads.

100. Hairy Mussel/Purewha
Modiolus areolatus
(Family Mytilidae)

Environment: Beneath rocks at low tide and in deeper water. *Range*: New Zealand. *Colour*: Exterior: Chestnut brown. Interior: White, with a reddish-purple area near the hinge. *Occurrence*: Fairly common. *Anterior – posterior*: 44-101mm. *Dorso – ventral*: 29-58mm. *REMARKS*: Takes its name from the stiff hair-like tufts found on the live shell at its posterior end. It also houses among the tufts other small plants and animals, helping to disguise the mussel.

101. Small Black Mussel/Hānea
Xenostrobus pulex
(Family Mytilidae)
Other Māori name: Kukupara

Environment: Middle and upper tidal rocky shore. *Range*: New Zealand. *Colour*: Exterior: Shiny blue-black. Interior: Bluish. *Occurrence*: Very common. *Anterior – posterior*: 22-30.5mm. *Dorso – ventral*: 11.5 – 13.5mm. *REMARKS*: A very similar, but larger species, *X. securis*, is found in brackish water, near the mouths of streams. Both species are also found in Australian waters. "Pulex" means "flea" hence the alternate name "flea mussel".

102. Horse Mussel/Hururoa
Atrina zelandica
(Family Pinnidae)
Other Māori names: -
Hoe moana, Kūkukuroa, Waharoa

Environment: Buried in mud at and below low tide. *Range*: New Zealand. *Colour*: Exterior: Young specimens – Light grey. Large specimens – Grey-black. Interior: Iridescent metallic colour. *Occurrence*: Fairly common. *Length*: 258-300mm. *Width*: 110-120mm. *REMARKS*: The largest and most fragile bivalve found in New Zealand waters. Buries itself point first in sand and mud. It protrudes only 12mm above the beach in very shallow water, but in deeper water more of the shell will protrude. The shell is decorated with radiating ridges and hollows.

103. Queen Scallop/Pīwhara
Pecten novaezelandiae
(Family Pectinidae)
Other Māori names: Kopa, Kuakua, Piwara, Tipa, Tupa

Environment: In sand at and below low tide. *Range*: North Island and northern part of South Island. *Colour*: Many shades of purple, brown, pink and white in variable patterns. *Occurrence*: Common. *Height*: 75-132mm. *Width*: 86-157mm.
REMARKS: A heavily ribbed, or "scalloped" shell. This was used by the Māori in pre-European times to cut the hair and to make necklace ornaments, *hei pīwhara*, that were much cherished. A favourite shellfish food of both Māori and Pākehā.

104. Fan Scallop/Kopakopa
Talochlamys zelandiae
(Family Pectinidae)
Other Māori names: Tipa, Tupa

Environment: Under rocks and among coral at and below low tide. *Range*: North Island and northern part of South Island. *Colour*: Many shades of purple, mauve, red, orange and rarely yellow, in variable patterns. *Occurrence*: Common. *Height*: 26-35mm. *Width*: 23-32mm.
REMARKS: One of the most brightly coloured and attractive of N. Z.'s shells.

105. Southern Fan Scallop
Zygochlamys delicatula
(Family Pectinidae)

Environment: In sponges, shallow to deeper water. *Range*: South Island. *Colour*: Many shades of purple, mauve, orange, red and yellow, usually with little or no pattern, but with darker zones. *Occurrence*: Fairly common. *Height*: 55.5mm. *Width*: 54.5mm. ***REMARKS***: Very similar to *Talochlamys zelandiae*, but even larger and more brightly coloured.

106. Deep Water Fan Scallop
Talochlamys gemmulata
(Family Pectinidae)

Environment: Deep water. *Range*: New Zealand except Northland and Southland. *Colour*: Many shades of pink, orange and white in variable patterns. Rarely pure white. *Occurrence*: Uncommon. *Height*: 31.5-65mm. *Width*: 28.5-64mm.
REMARKS: This species is sometimes found in cod stomachs. It is replaced in the north by *T. consociata*.

107. Purple Fan Scallop
Talochlamys gemmulata radiata
(Family Pectinidae)

Environment: Deep water. ***Range***: Fouveaux Strait and Stewart Island. ***Colour***: Usually deep rich purple, also shades of orange, pink, brown and white. ***Occurrence***: Uncommon. ***Height***: 55-75mm. ***Width***: 55-75mm. ***REMARKS***: The scallop is a filter-feeding mollusc. The animal lives in a slight depression on the sea bottom. It moves about by closing its shell rapidly and shooting out a jet of water, thus thrusting itself forward or backwards.

108. Striped Fan Scallop
Talochlamys zeelandona
(Family Pectinidae)

Environment: Low tide, under rocks, to deeper water. ***Range***: North Island and northern part of South Island. ***Colour***: Variable. Ranges through white, yellow, orange, pink, red to deep purple. Often boldly striped. ***Occurrence***: Rare. ***Height***: 26-35mm. ***Width***: 23-32mm. ***REMARKS***: Unlike other scallops who move about their habitats, *C. zeelandona* lives fixed permanently to the undersides of rocks, anchored by threads (byssus) proceeding from the foot of the animal.

109. Lion's Paw
Mesopeplum convexum
(Family Pectinidae)

Environment: Deep water. ***Range***: New Zealand. ***Colour***: Shades of pink, purple, orange, white and very rarely yellow, in variable patterns. Interior: Pinkish or white. ***Occurrence***: Rare. ***Height***: 45-57mm. ***Width***: 48-62mm. ***REMARKS***: The hinges are unequal, as is the case with other scallops. This is another scallop sometimes found in the stomachs of cod. Also occasionally washed ashore in storms.

110. File Shell
Lima zealandica
(Family Limidae)

Environment: From shallow to deep water. ***Range***: South Island. ***Colour***: Exterior: White to dull reddish-brown, but often heavily encrusted. The eighteen ribs are sometimes tinted with brown or green. Interior: White, polished, the ribs distinctly visible. ***Occurrence***: Uncommon. ***Height***: 38-68mm. ***Width***: 26.5-53mm

111. Little File Shell
Limatula maoria
(Family Limidae)

Environment: In sand, shallow to deep water. ***Range***: New Zealand. ***Colour***: White. ***Occurrence***: Uncommon. ***Height***: 20-29mm. ***Width***: 12.5-16.8mm.
REMARKS: This shell has 25 or 26 low radial ribs that are like those on a closely spaced file tool, hence the name. The animal has tentacles, but no eyes, yet moves about easily in its sandy habitat.

112. Golden Oyster/Poro
Anomia trigonopsis
(Family Anomiidae)

Environment: Low tidal rocks. ***Range***: North Island. ***Colour***: Variable. Silvery-white and translucent. Also sea-green when living, yellow to bright orange when polished by long wave movement. ***Occurrence***: Fairly common. ***Height***: 40-67mm. ***Width***: 51-83mm. ***REMARKS***: Used in craftwork to make petals. Also known as a jingle shell. A close relation, the Window Oyster, *Pododesmus zelandicus*, is commoner in the south.

113. Stewart Island Oyster/ Tio para *Ostrea chilensis*
(Family Ostreidae)
Other Māori name: Tio pati

Environment: Low tide to deep water, rocky and muddy areas. ***Range***: New Zealand, particularly Fouveaux Strait. ***Colour***: Exterior: Rough dull grey. Interior: White, stained with olive green. ***Occurrence***: Common. ***Height***: 74-105mm. ***Width***: 63-70mm. ***REMARKS***: Has been taken from Fouveaux Strait commercially since the 1880's. Trawled there in deep water during a limited season.

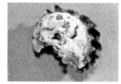

114. Auckland Rock Oyster/ Tio repe
Saccostrea glomerata
(Family Ostreidae) Other Māori names: Karauria, Ngākihi, Repe, Tio, Tio kōhatu, Tio reperepe

Environment: On low tidal rocks. ***Range***: Northern part of North Island. ***Colour***: Exterior: Slate grey, edged with dark purple. Interior: white with purplish blotches. ***Occurrence***: Very common. ***Height***: 60-90mm. ***Width***: 35-68mm. ***REMARKS***: Also farmed commercially. Farmed by the Māori in earlier times, see the introduction. Much of the commercial farming in the north is now of the Pacific oyster, *Crassostrea gigas*, which was introduced to the country accidentally in the 1970's.

115. Nestling Cockle
Cardita aoteana
(Family Carditidae)

Environment: Clustering under rocks at extreme low tide and in deeper water. *Range*: New Zealand, except Auckland. *Colour*: Yellowish-brown to chestnut. *Occurrence*: Uncommon. *Anterior – posterior*: 28-35.5mm. *Dorso – ventral*: 16-25.5mm. *REMARKS*: A solid shell with prominent ribs. The Northland species, *C. brookesi*, is a much smaller shell that is usually fawn to pale brown in colour.

116. Purple Cockle/Pūrimu
Purpurocardia purpurata
(Family Carditidae)

Environment: In sand in shallow water. *Range*: New Zealand. *Colour*: Exterior: Mottled pale brown and white if not corroded by the waves; variable otherwise. Interior: White, or white flushed with pink. *Occurrence*: Common, especially in the north. *Height*: 31-47mm. *Width*: 35-50mm. *REMARKS*: Is found living attached to kelp roots that usually discolour one end of the shell. Was collected in past times by the Māori in baskets called *hau* or *rori*. A more elongated species, *P. reinga*, is found at Cape Maria van Dieman.

117. Lace Cockle
Divaricella huttoniana
(Family Lucinidae)

Environment: In sand at and below low tide. *Range*: New Zealand. *Colour*: White. *Occurrence*: Fairly common. *Height*: 26-40mm. *Width*: 26.75-43mm. *REMARKS*: Easily recognised by its sculpture of diverging lines. In Māori mythology all cockles are said to originate from their progenitors Te Arawaru and Kaumaihi. Pipihura is also said to have produced cockle species.

118. Strawberry Cockle
Pratulum pulchellum
(Family Cardiidae)

Environment: Below the tide, sandy ground. *Range*: New Zealand. *Colour*: Exterior: Pinkish with grey-brown epidermis. Interior: White, flushed with pink or red. *Occurrence*: Not common. *Height*: 14.5-26mm. *Length*: 16-29mm. *REMARKS*: This is the only true cockle found in New Zealand waters, although the word "cockle" is used for a number of other bivalves. An attractive shell in form and colour with fine rib sculpture.

119. Silky Dosinia/
Tupa *Dosinia lambata*
(Family Veneridae)

Environment: In mud below low tide. *Range*: North Island and northern part of South Island. *Colour*: Exterior: White, often with pale yellow umbones (points or beaks). Interior: White, polished. *Occurrence*: Common. *Height*: 22.5-24mm. *Width*: 24-35mm. *REMARKS*: A fragile shell with very fine sculpturing. The rare *D. maoriana* is very similar in size and shape, but has a much thicker shell, and is dull white externally.

120. Ringed Dosinia/
Tipatipa *Dosinia anus*
(Family Veneridae)
Other Māori names: -
Harihari, Pātiki, Tuanga haruru

Environment: In sand at low tide. *Range*: North Island and northern part of South Island. *Colour*: Exterior: Light yellowish or reddish-brown. Interior: White, flushed with mauve, porcellanous. *Occurrence*: Common. *Height*: 55-78mm. *Width*: 58.5-80.5mm. *REMARKS*: Also called the biscuit shell. Is heavily sculptured with circular ridges; "anus" in the classification name means "ring".

121. Fine Dosinia/
Hahari *Dosinia subrosea*
(Family Veneridae)
Other Māori names: -
Hākari, Harihari

Environment: In sand at low tide. *Range*: North Island and northern part of South Island. *Colour*: Exterior: Beige or white. Interior: Dull white. *Occurrence*: Fairly common. *Height*: 41-52mm. *Width*: 44.5-56.5mm. *REMARKS*: A smooth shell to touch. The sculpturing is much finer than *D. anus*, but not as fine as *D. lambata*.

122. Deep Water Venus Shell
Notocallista multistriata
(Family Veneridae)

Environment: Deep water. *Range*: New Zealand. *Colour*: Exterior: Reddish-brown and white in variable patterns. Interior: White, lightly flushed with purple. *Occurrence*: Rare. *Height*: 18.5-25.5mm. *Length*: 25-33mm. *REMARKS*: Has a polished surface and closely spaced concentric lines. This shell is occasionally found in fish stomachs.

123. Coarse Dosina
Dosina zelandica
(Family Veneridae)

Environment: Muddy ground; low tide to deeper water. ***Range***: New Zealand. ***Colour***: Exterior: Light brown, sometimes with brown markings. Heavily sculptured with ridges. Interior: White- porcellaneous. ***Occurrence***: Common. ***Height***: 39-51mm. ***Length***: 48.5-66mm.

125. Frilled Venus Shell/ Pūkauri *Bassina yatei*
(Family Veneridae)

Environment: Below low tide, in sand. ***Range***: North Island and northern part of South Island. ***Colour***: Exterior: Creamy white with light mauve umbones. Interior: White. ***Height***: 45-54mm. ***Length***: 53.5-63.5mm. ***REMARKS***: A beautiful shell with delicate frills. These frills are very high on young shells, but as they attain full size, the frills are worn down by the movement of the waves over the sand.

124. Morning Star/ Tāwera *Tawera spissa*
(Family Veneridae)

Environment: In sand, low tide to deep water. ***Range***: New Zealand. ***Colour***: Exterior: White with an astonishing variety of zigzag markings in dark pink or brown, or with heavy bands of similar colours. Interior: White, flushed with purple. ***Occurrence***: Common. ***Height***: 15.75-24mm. ***Length***: 20.25-28.5mm.
REMARKS: Tāwera is the Māori name for the morning star, the planet Venus.

126. Cockle/Tuangi
Austrovenus stutchburyi
(Family Veneridae)

Other Māori names: - Anga, Hinangi, Hūai, Huangi, Hūngangi, Hūwai, Pipi tuangi, Tanetane, Tuaki, Tungangi

Environment: On mudflats, intertidal. ***Range***: New Zealand. ***Colour***: Juveniles sometimes have bright yellow colouring.

Exterior: Whitish. Interior: Whitish, deeply flushed with purple. *Occurrence*: Very common. *Height*: 47-56mm. *Length*: 51-62mm. *REMARKS*: Like many other *Veneridae*, not a true cockle in spite of its name. A major food source. A cutting reference to any people who will not be aroused to avenge an injury is: *Tēnei te iwi te takoto tonu nei me he moe tuangi!*: "These are the folk who lie down to it like the sleeping tuangi!" A similar phrase is quoted concerning the toitoi or kāeo (Cook's Turban, No.29).

127. Ribbed Venus Shell/ Kaikaikaroro
Protothaca crassicosta
(Family Veneridae)
Other Māori names: -
Karoro, Tuaki, Tuangi

Environment: In sand or mud among rocks. *Range*: New Zealand. *Colour*: Exterior: Creamy white. Interior: White flushed with violet. *Occurrence*: Common. *Height*: 31-35.5mm. *Length*: 36.5-45mm. *REMARKS*: A solid shell with broad radial ribs and concentric grooves, the whole surface being divided into three distinct radial zones.

128. Oblong Venus Shell/ Hahari
Ruditapes largillierti
(Family Veneridae)
Other Māori names: -
Hākari, Harihari, Tūaki

Environment: In sand or mud at and below low tide. *Range*: New Zealand. *Colour*: Exterior: Whitish, sometimes lightly patterned with brown. Juveniles often have grey-brown lines. Interior: White, sometimes flushed with purple. *Occurrence*: Very common. *Height*: 33-56mm. *Length*: 40-70mm. *REMARKS*: A quite small shell with a polished surface and very fine sculpture.

129. Angled Wedge Shell/ Hohehohe
Peronaea gaimardi
(Family Tellinidae) Other Māori names: Kaikaikaroro, Kuharu, Kuwharu, Pipi peraro, Pipi toretore, Peraro, Taiwhitiwhiti

Environment: In sand in shallow water. *Range*: New Zealand. *Colour*: Exterior: Semi-translucent white. Interior: Bluish-white. O*ccurrence*: Common. *Height*: 38-43mm. *Length*: 58-65mm. *REMARKS*: A triangular-shaped shell with fine and dense concentric lines. *Tellinota edgari* is very similar, but it is a little smaller, and flushed with yellow.

130. Spencer's Wedge Shell
Rexithaerus spenceri
(Family Tellinidae)

Environment: In sand below low tide. *Range*: New Zealand. *Colour*: Exterior: White. Interior: White, porcellanous. *Occurrence*: Rare. *Height*: 17-31.5mm. *Length*: 29-56mm. *REMARKS*: Elongated shell, very pointed at one end and rounded at the other. Its epidermis is very thin, yellowish, and therefore easily rubbed off.

131. Large Wedge Shell/ Hanikura
Macomona liliana
(Family Tellinidae)

Other Māori name: Hanikura patu

Environment: In sandy mud at low tide. *Range*: New Zealand. *Colour*: Exterior: creamy white, semi-translucent. Interior: Shiny white, diffused by yellow. *Occurrence*: Common. *Height*: 30-50.5mm. *Length*: 36.5-66mm. *REMARKS*: Very large specimens have been found at Cheltenham Beach, Auckland where they are protected from severe weather. Also called the tulip or lily shell.

132. Round Wedge Shell
Pseudacopagia disculus
(Family Tellinidae)

Environment: In sandy mud at low tide. *Range*: New Zealand. *Colour*: Dull creamy white, deeply flushed with yellow around the umbones, particularly bright yellow inside the shell. *Occurrence*: Common. *Height*: 23-34mm. *Length*: 29-39mm. *REMARKS*: Has fine concentric rings and a thin epidermis that is persistent only near the rims.

133. Purple Sunset Shell/ Kuharu *Gari stangeri*
(Family Psammobiidae)
Other Māori names: -
Kuwharu, Ururoa, Wahawaha

Environment: In sand at low tide. *Range*: New Zealand. *Colour*: Exterior: White, rayed with various shades of violet or purple. Interior: Violet or purple. *Occurrence*: Common. *Height*: 25.5-34mm. *Length*: 42-58mm. *REMARKS*: A thin and fragile shell, but even so the largest and least fragile of the sunset shells we list.

134. Pink Sunset Shell/
Takarape *Gari lineolata*
(Family Psammobiidae)
Other Māori name: Takarepo

Environment: In sand at low tide. ***Range***: New Zealand. ***Colour***: Exterior: Pink with shades of deeper pink in concentric bands. Interior: Pinkish-purple. ***Occurrence***: Common. ***Height***: 27-41mm. ***Length***: 53-80mm. ***REMARKS***: The shell is narrowly ovate with a distinct ridge at the hinge. The name "lineolata" has the meaning "marked with lines".

136. Large Trough Shell/
Kuhakuha *Mactra discors*
(Family Mactridae)
Other Māori names: Kaikaikaroro, Whāngai karoro, Whakai o tama

Environment: In sand at low tide. ***Range***: New Zealand. ***Colour***: Light brownish-white, edged with khaki epidermis. ***Occurrence***: Common. ***Height***: 65-80mm. ***Length***: 76-92mm. ***REMARKS***: Living specimens are driven ashore during heavy gales; good eating. Sometimes yellowish when newly stranded. The word "discors" means "different"

135. Shining Sunset Shell/
Pīpipi *Soletellina nitida*
(Family Psammobiidae)

Environment: In sand at low tide. ***Range***: New Zealand. ***Colour***: Violet, inside and out, but the outside mostly covered with a yellowish-brown epidermis. ***Occurrence***: Common. ***Height***: 16-23.5mm. ***Length***: 33-46mm. ***REMARKS***: A very fragile shell being both thin and almost transparent. Glossy, horn-like epidermis. *S. siliquens* is very similar, but is white instead of violet.

137. Oval Trough Shell/
Ruheruhe
Cyclomactra ovata
(Family Mactridae)
Other Māori name: Tuaki

Environment: Muddy areas. ***Range***: New Zealand. ***Colour***: Exterior: White to yellowish-brown. Interior: Yellowish on the upper parts, bluish-white towards the base. ***Occurrence***: Fairly common. ***Height***: 54-66mm. ***Length***: 63-85mm. ***REMARKS***: *C. tristis* is very similar to *C. ovata*, but is more elongated.

138. Elongated Mactra/ Poua *Oxyperus elongata*
(Family Mactridae)
Other Māori names: Poue, Roroa, Tohemanga, Tupehokura

Environment: In sand below low tide. *Range*: New Zealand. *Colour*: Exterior: White mottled with pale chestnut. Sometimes has darker distinct bands, dots and splashes. Interior: Yellowish, porcellanous. *Occurrence*: Uncommon. *Height*: 56-67mm. *Length*: 88.5-111mm. *REMARKS*: A heavier shell than other *Mactra* with fine concentric rings over most of the shell.

139. Triangle Shell/Kaikaikaroro *Spisula aequilatera*
(Family Mactridae) Other Māori names: Kaiparoro, Whakai o tama

Environment: In sand at low tide. *Range*: New Zealand. *Colour*: Exterior: Adults – Bluish-white, occasionally brown. Juveniles – Coloured with white, particularly inside. Usually has a bluish-violet patch around the hinge. Interior: Yellowish-white, porcellanous, polished. *Occurrence*: Common. *Height*: 47-60mm. *Length*: 56-77mm. *REMARKS*: The name "*aequilatera*" means "having equal sides".

140. Scimitar Mactra/ Peraro *Zenatia acinaces*
(Family Mactridae)
Other Māori name: Pipi roa

Environment: Buried in sand below low tide. *Range*: New Zealand. *Colour*: Exterior: White covered with yellowish-brown periostracum. Interior: Bluish-green, pearly and iridescent. *Occurrence*: Not common. *Height*: 18-48mm. *Length*: 47-101mm. *REMARKS*: A rather fragile shell. The beak is markedly off-centre, whereas that of *R. lanceolata*, below, is near the centre.

141. Lance Mactra/ Pipi rahi *Resania lanceolata*
(Family Mactridae)
Other Māori names: Pipi, Roroa

Environment: Buried in sand below tide. *Range*: New Zealand. *Colour*: Exterior: Smooth creamy white with a thin, pale chestnut, epidermis. Interior: White, lightly iridescent. *Occurrence*: Not common. *Height*: 42-49mm. *Length*: 93-100mm. *REMARKS*: Another rather fragile shell, "*lanceolata*" means "in the shape of a lance".

142. Pipi
Paphies australis
(Family Mesodesmatidae)
Other Māori names: Angarite, Anutai, Hinangi, Kahitua, Kākahi, Kōkota, Ngaingai, Pīpihi, Pipitaiawa, Roroa, Taiwhatiwhati, Tuaki

Environment: Low tide in mud or sand. *Range*: New Zealand. *Colour*: Whitish with cream epidermis. Sometimes stained black due to immersion in mud. *Occurrence*: Very common. *Height*: 30.5-51mm. *Length*: 48-83mm. **REMARKS**: Like its cousins the toheroa and tuatua, this shellfish is very good to eat. If not cooked at once in former times the Māori sun-dried them and threaded them on strings for later use.

Another past use was to give water from cooked pipi as a tonic to invalids, and women who had just had a baby. Also taken for coughs and sore throats. Several proverbs mention the value of this broth: *Ka whakangotea ki te wai o te kākahi*, "It was suckled on the juice of the kākahi", is a reference to someone who has grown to manhood strong because of it. Similarly: *Ko te kākahi te whaea o te tamaiti*, "The kākahi is the mother of the child". Then this local saying: *Iti koe e ngā pipi o Hokianga, he wai ū tangata tonu*. "Little though you are, the pipi of Hokianga, you are like mother's milk. That is to say, something small can be of great value.

Another local saying is: *Haere ki te kōwhā pipi ki Katikati*, or "Go to Kaitikati to shell pipi". A jibe implying it is not safe to do something. Which reminds of the old belief that menstruating women were not to go down the beach to collect pipi for fear that the shellfish would move to another part of the beach. Then there is the description *Me te kōkota*, "As fair as the ivory-shelled pipi", a term of endearment. On the other hand *Me te niho kōkota*, "Like kōkota teeth", was said of enemy heads put on stakes and jeered at by the people.

Pipi shells were used for scaling fish, in incising the gums of children, in scraping flax when making garments, in shaving the hair, in circumcision, and as tweezers to remove hair. Pipi shells were put on a fire of mānuka sticks to keep the fire continually heated. Quite as numerous uses as the proverb portends: "There they lie as numerous as pipi" (*Tēnā, tērā te noho ana me te one pīpihi*). A saying for the great numbers of people living in a single settlement.

143. Toheroa
Paphies ventricosa
(Family Mesodesmatidae) Other Māori names: Moeone, Toimanga

Environment: Low tide, in sand. *Range*: New Zealand. *Colour*: Exterior: Creamy white. Interior: White. *Occurrence*: Common but local. *Height*: 62-81mm. *Length*: 97-117mm. **REMARKS**: Toheroa means "long tongue" from the protruding fleshy ligament with which the animal rapidly works its way below the surface of hard wet sands on the seashore. The taking of this shellfish is now very restricted due to its great popularity. It is said that in the early 1900's toheroa were so plentiful in

Northland that horses and ploughs were taken to the beaches to plough up the shellfish like potatoes. Mareao, the ancestor of the Ngāti Whatua of Northland, is claimed to have brought the toheroa from legendary Hawaiki, but others would dispute that.

144. Tuatua
Paphies subtriangulata
(Family Mesodesmatidae)
Other Māori names: Kahutua, Kaitua, Pipi tairaki, Tairaki, Taiwhatiwhati

Environment: In sand at low tide. *Range*: North Island and northern part of South Island. *Colour*: White under yellowish periostracum. *Occurrence*: Common. *Height*: 31.5-48mm. *Length*: 48.5-76mm. *REMARKS*: It is difficult to distinguish a young toheroa from an adult tuatua, though the latter are much more plentiful. Both shell shapes are similar but the toheroa has a thinner shell than the tuatua. Pipi and tuatua are also much confused. The pipi being more rounded in shape, compared to the more triangular shape of the tuatua.

145. Deep Burrower/ Hohehohe
Panopea zelandica
(Family Hiatellidae)

Environment: Deeply buried in sand below low tide. *Range*: New Zealand. *Colour*: Exterior: White, faintly tinged with pale orange-brown. Interior: White. *Occurrence*: Uncommon. *Height*: 37.5-68.5mm. *Width*: 65-121mm. *REMARKS*: A fragile shell with gaping ends. *P. smithae*, a deep water species from Fouveaux Strait, is pure white with a much more solid shell.

146. Rock Borer/Pātiotio
Barnea similis
(Family Pholadidae)
Other Māori name: Tetere moana

Environment: Boring into soft rock at low tide. *Range*: New Zealand. *Colour*: Exterior: Whitish. Interior: White with concentric ribs. *Occurrence*: Fairly common. *Height*: 22-32mm. *Length*: 68-86mm. *REMARKS*: The valves gape widely at the lower end, unlike those of its smaller relatives *Philadidae suteri* and *P. tridens*, which are closed. When collecting *B. similis* alive, watch out for the tiny bivalve *Arthritica crassiformis* that lives in the Rock Borer's burrows.

DEEP SEA CEPHALOPO[DS]

147. Battleaxe/
Pākira *Myadora striata*
(Family Myochamidae)
Other Māori names: Pukira, Toitoi

Environment: In sand at and below low tide. *Range*: New Zealand. *Colour*: Exterior: White with weak concentric ridges. Interior: Shining white. *Occurrence*: Not common. *Height*: 30-36mm. *Length*: 35-42.5mm. *REMARKS*: This shell has a unique shape, described as ovate-triangular. One valve is quite flat and the other slightly concave.

148. Lantern Shell
Offadesma angasi
(Family Periplomatidae)

Environment: In mud below low tide. *Range*: North Island and northern part of South Island. *Colour*: Exterior: White, or dirty white. Interior: White, lightly nacreous. *Occurrence*: Rare. *Height*: 53-65.6mm. *Length*: 79-94mm. *REMARKS*: A very fragile shell, with an oddly shaped hinge. One valve is much flatter than the other, as was the case with the previous shell listed, the Battleaxe.

149. Paper Nautilus/
Muheke *Argonauta nodosa*
(Family Argonautadae)
Other Māori names: -
Ngū (egg case), Pūpū tarakihi

Environment: Inhabiting the open ocean. *Range*: New Zealand. *Colour*: White with brown on the keels. *Occurrence*: Rare, occ. large numbers are washed ashore. *Width*: Up to 304mm. *REMARKS*: This extremely fragile and beautiful shell nevertheless helps the female protect her eggs. The animal has no muscles to attach itself to its shell. Hence, when it is in danger it swims to safety. The male is much smaller and without a shell.

150. Ram's Horn/Kotakota ngū
Spirula spirula
(Family Spirulidae)

Environment: Deep water. *Range*: New Zealand. *Colour*: Translucent white. *Occurrence*: Common. *Width*: About 20mm. *Remarks*: Although this fragile little "shell" is often washed ashore in large numbers, the living animal, a small deep-water squid with a fawn body, is rarely seen. Octopi are its close relative. The "shell" is in fact part of its internal skeleton. If in danger the animal can withdraw within its mantle cavity which then closes. It propels itself along with two small end fins.